GLOBAL
DISCONTENTS

Noa⁀ Chomsky is the author of numerous bestselling political works, inc⁀⁀ing *Hegemony or Survival* and *Failed States*. A laureate professor at the ⁀niversity of Arizona and professor emeritus of linguistics and phil⁀⁀ophy at MIT, he is widely credited with having revolutionized mode⁀n linguistics. He lives in Tucson, Arizona.

Davi⁀ Barsamian is the award-winning founder and director of *Alterna-tive ⁀⁀io*, an independent syndicated radio program. In addition to his ten b⁀⁀ks with Noam Chomsky, his works include books with Tariq Ali, Howa⁀d Zinn, Edward Said, Arundhati Roy and Richard Wolff. He lives in Bo⁀⁀der, Colorado.

NOAM
CHOMSKY

GLOBAL DISCONTENTS

CONVERSATIONS ON
THE RISING THREATS
TO DEMOCRACY

INTERVIEWS WITH

DAVID BARSAMIAN

PENGUIN BOOKS

PENGUIN BOOKS

UK | USA | Canada | Ireland | Australia
India | New Zealand | South Africa

Penguin Books is part of the Penguin Random House group of companies
whose addresses can be found at global.penguinrandomhouse.com.

First published in the United States of America by
Metropolitan Books, Henry Holt and Company 2017
First published in Great Britain by Hamish Hamilton 2017
Published in Penguin Books 2018

001

Special thanks to Anthony Arnove, Kadriye Barsamian, Elaine Bernard,
Sara Bershtel, Glenn Ketterie, Joe Richey, Sophie Siebert,
Bev Stohl, and Abha and Mriganka Sur.

Printed in Great Britain by Clays Ltd, St Ives plc

A CIP catalogue record for this book is available from the British Library

ISBN: 978-0-241-98199-3

www.greenpenguin.co.uk

Penguin Random House is committed to a
sustainable future for our business, our readers
and our planet. This book is made from Forest
Stewardship Council® certified paper.

CONTENTS

1

STATE SPYING AND
DEMOCRACY

Edward Snowden's revelations of widespread state surveillance of Internet and telephone communications have caused some consternation here in the United States—and around the world. Were you at all surprised by the government's electronic dragnet?

Somewhat—not a lot. I think we can take for granted that if technology or other means of control and domination are available, then power systems are going to use them. Take the recent revelations about the relationship between the National Security Agency (NSA) and Silicon Valley. Silicon Valley is a synonym for the commercial use of surveillance. The NSA is going to Silicon Valley for help, because the commercial enterprises have been doing this already, on a great scale, and they have the technological expertise. So apparently, a private security officer was brought to the NSA to help them develop sophisticated techniques of surveillance and control.[1]

The technology is available. You can use it for making money, and you can use it for controlling people's attitudes and beliefs, directing them toward what you want them to do. So they do.

In fact, anyone who has paid any attention to history should not be in the least surprised. Go back a century to the U.S. war in the Philippines. The United States invaded the Philippines, killed hundreds of thousands of people, and finally suppressed the resistance. But then they had to pacify the population. There are very good studies about this, particularly by Al McCoy, a Philippines historian. He shows that the U.S. was very successful in pacifying the population, using the most advanced intelligence-gathering and surveillance techniques of the day—not our technology but what you had a century ago—to sow distrust, confusion, and antagonisms, standard devices of counterinsurgency. He also points out that it was just a few years before these techniques were used back home. The Woodrow Wilson administration used them in their Red Scare. That's how it works.[2]

Just like with drones. Another recent admission, hardly a surprise, is that the Federal Bureau of Investigation (FBI) has been using drones for surveillance. You use them against those you designate as enemies, and you very quickly adapt the same technology at home. And there's more to come. For years, the military and the security system in general have been trying to develop fly-sized drones, which can get into your living room to see and record everything that's happening. The robotics labs have gotten to the point where they're about ready to deliver this technology.

If you look abroad, drones were at first used for surveillance. Later, they were used for murder. And we can expect that domestically, too. If there is a suspect, maybe a person with the wrong idea, maybe someone like Fred Hampton, instead of sending the Chicago police in to murder him, maybe you can murder him with a drone. We can expect that.

Fred Hampton being the Black Panther activist in Chicago who was murdered, along with another Black Panther, Mark Clark, in 1969.

That was a real Gestapo-style assassination, which stayed undercover for the longest time.

If a technology is available, a system of power is going to make use of it. Back through history, that's what you find. So to be surprised is to have blind faith in power systems that somehow they're not going to use what's available to them as a way of controlling, dominating, and indoctrinating people. But, of course, whether commercial enterprises or a state, they will. Yes, the particulars can be surprising. I didn't know that they had the PRISM program, a secret tool that allowed NSA officials to gather search histories, e-mails, chats, and other data directly from companies like Google and Facebook.[3] But you can't really be surprised at the general phenomenon.

To take just one more example: the *MIT Technology Review* had a news item describing how corporations are becoming wary about using computers with components manufactured in China, because apparently it's technically possible to design hardware that will detect everything the computer is doing.[4] Naturally, they don't add the next point, which is that if the

Chinese can do it, the United States can do it much better. So what's so safe about using computers with hardware manufactured in the United States? Pretty soon, we'll have every keystroke sent off to the president's database in Utah.

You've noted that there is a generational gap in responses to these NSA revelations.

I haven't seen a study, but what I sense—at least, from discussion and what I've read—is that younger people are less offended by this than older people. And I suspect it's part of a cultural shift that's taking place, among younger people particularly, toward a kind of exhibitionist culture. You put everything on your Facebook page: what you're doing, what you're wearing, what you're thinking. Everything is exposed. And if everything is exposed, who cares if the government sees it?

Do you see this trend toward a surveillance state as part of a drift toward totalitarianism? Or is that too strong a term?

It's a move in that direction. But there is a considerable gap between collecting data and having a way to use it. One of the more positive aspects of this, if you want to put it that way, is that the authorities probably don't have the competence to make use of the material they are collecting. They can make use of it for particular purposes. If there's this huge database in Utah, which is going to have massive information on everybody sooner or later, if there's some person they want to go after—the next Fred Hampton, let's say—then they can get plenty of information about that person, and it may enable them to control or maybe even kill that person. But short of that, it's not clear that they can do very much.

We saw this in the past. The FBI, using much more primitive means, had tons of data about everyone. We all knew back in the 1960s, and ever since, that every activist organization was probably infiltrated by government spies. In fact, people pretty quickly learned that if you wanted to do anything sensitive, you did it with an affinity group, not even with your comrades, because one of them was probably a police informant. But in many ways, the government couldn't do much with the data they gathered. They could do certain things, like harming a particular individual. But if you look at, say, the trials of the Vietnam War resistance, it was amazing what the FBI couldn't do.

I followed those trials closely.[5] The main one was the so-called Spock-Coffin trial.

That involved Dr. Benjamin Spock and Reverend William Sloane Coffin Jr., who were accused of conspiring to help young men resist the draft.

I was an unindicted co-conspirator, so I sat in on the trial. After the prosecution rested its case, the defense met to decide what to do. The original thought had been, It's an open-and-shut case, so just say that everyone is guilty. Don't deny it. In fact, proudly proclaim it. Put on a political defense. But the defense lawyers decided not to put on a defense at all, because the prosecution's case turned out to be really weak. People were standing up at Town Hall in New York and saying, "We hereby conspire to undermine the Selective Service Act"—but the government just didn't pay attention to that. Instead, they spent all their time trying to figure out where we were getting our instructions from. Was it Hungary or North Korea? What were we really doing? It couldn't be what was said in the open. So they just missed everything.

It was the same with the Pentagon Papers. When Dan Ellsberg was underground—he hadn't yet revealed himself—a number of people were distributing the papers. I was one of them. The press was after me all the time. I was getting regular calls from newspapers from the United States and abroad asking to see part of the Pentagon Papers. But the FBI never figured out that I had them. After Dan surfaced and identified himself, then FBI agents came to my house to question me. *After.* Apparently, they just hadn't been able to figure out what the press had figured out.

There's case after case like this. Their own mentality directs them to certain kinds of conspiratorial actions, yet a lot of resistance is purposely open. You're trying to reach people, explain to them what you're doing. It's not underground. Some things, like getting a deserter out of the country, you do quietly. But other things—like saying, "Let's refuse to pay taxes" or "Let's break down this legal system, which is causing vast atrocities and crimes"—those aren't hidden.

When Obama was first elected, you were not part of the chorus of cheerleaders. What about the continuities between George W. Bush and Barack Obama? Are there any?

Yes, there are real continuities. Obama extended enormously some of the more harmful and, in fact, criminal aspects of Bush's programs.

Obama is credited with having withdrawn troops from Iraq and Afghanistan. But those withdrawals were already under way. In Iraq, it was clear that the United States was basically defeated. Its war aims were unrealizable, and the Bush administration was starting to pull out. In Afghanistan, Obama actually expanded the war in the hope of achieving some kind of victory. It didn't happen. There, too, the troops were going to have to leave. So those withdrawals were nothing special.

But he extended other programs, like the drone program. Right away. And we should remember that this is an international terrorist campaign—the world's leading international terrorist campaign. If you're living in a village in Yemen or North Waziristan and you don't know whether in five minutes a sudden explosion across the street will blow away a bunch of people who are standing there, and maybe hit you as well, you're terrorized. You live with this fear all the time. That's sheer terror, by the narrowest definition of the term. And this is happening on a massive scale.

There's a lot of focus on so-called signature strikes, in which we don't actually know who the person is that we're shooting at. We're killing people just because their age, location, and behavior supposedly match the "signature" of terrorist activity. And that's clearly bad. But the whole idea of drone strikes is outrageous. It's pure terrorism on a scale that Al Qaeda couldn't dream of.

Furthermore, this campaign is generating terrorists—and is known to be doing so. The highest-level officials and commentators have pointed out that these attacks are creating potential terrorists. It's perfectly obvious why. A couple of days after the Boston Marathon bombings, for instance, there was a drone attack in Yemen. Usually we don't know much about these strikes. This one we do know about because a young man from the village happened to testify before the U.S. Senate. He said that for years the jihadis in Yemen had been trying to turn the village against the Americans, but it hadn't worked, because the villagers didn't know anything about the United States except what they'd been told by this young man, which was favorable. But after this one drone strike, he said, the whole village hates America.[6] And once people hate America, some will try to do something about it. So it's a terror-generating machine.

This also puts an interesting light on the discussion about the NSA exposures. The government justification for the surveillance is that we have to sacrifice some privacy for security. We're talking about a government that is deliberately undermining security—and creating a terrorist threat beyond any that exists. How can we not just collapse in ridicule when they say we have to have surveillance to promote security?

Actually, the idea that governments place a high priority on security is mythical. You learn it in international relations courses. But just take a look at history. It's easy to show that it's not true.

What is driving Obama's assault on civil liberties and prosecution of whistle-blowers?

That's a good question. As you mentioned, I really didn't expect very much from Obama. I wrote critically about him even before the primaries, just quoting his website. It was pretty clear that his campaign was smoke and mirrors. But I don't understand what he's gained by his enormous escalation of the attack on civil liberties.

He has prosecuted more whistle-blowers than all the presidents in the entire history of the country combined. But there are also other cases that the Obama administration has brought to the courts that involve major attacks on civil liberties. One of the worst is *Holder v. Humanitarian Law Project*. This is a legal-assistance group that was giving legal advice to the Kurdistan Workers' Party (PKK), a Kurdish group that is on the State Department's Foreign Terrorist Organizations list. Just giving legal advice. The Obama administration wanted to expand the notion of "material support for terrorism" to include advice. Material support used to mean handing you a gun. Now it means saying, "Here are your legal rights."

If you look at the discussion in the court proceeding, it's pretty clear that the administration interprets material support to mean almost any interaction with what they call terrorist groups.[7] So, say, if I meet Hassan Nasrallah, the head of Hezbollah, because I'm interested in knowing what he's doing—Nasrallah is an interesting person—that could be called material support for terrorism. That's a tremendous attack on civil liberties.

This is, incidentally, quite apart from the issue of the legitimacy of the State Department's terrorist list—which, unfortunately, is unquestioned. Why is it legitimate in the first place? Why is the state executive granted the authority to capriciously decide you're a terrorist? Why does the state have the right to say that Nelson Mandela is a terrorist, which they insisted on until just a couple of years ago?[8] Why do they have the right to say, as Ronald Reagan did in 1982, that Saddam Hussein isn't a terrorist, just because the U.S. government wanted to give him aid?[9] Can we even take this seriously? If somebody is put on a terrorist list, they have no recourse. There is no way of saying, "Look, I'm not a terrorist."

The government doesn't have to provide any evidence, and there's no judicial review. The list is an executive authorization for murder. We shouldn't accept this in the first place. And we shouldn't accept a concept of material support that says that if you tell someone their legal rights, or maybe even if you just have discussions with them, you're helping terrorism.

Given the structural constraints of the national security state, can a president fundamentally change U.S. foreign policy?

Sure. A president can't just say, "Okay, I'm going to change U.S. foreign policy." But the president has a lot of power to reach the public. Franklin Delano Roosevelt used this power. Lyndon Johnson used it. Public opinion can, I think, easily be turned against the national security state. If you look at polls, plenty of people are opposed to surveillance. The ones who support surveillance are the ones who are as deluded as people like Thomas Friedman or Bill Keller at the *New York Times*, who think that we have to have surveillance for the sake of security—not noticing that the very administration that is calling for defense against terrorism is maximizing terrorism and the threats against us. But these are things a president could overcome.

A president could say, "Look, these operations that we're carrying out are generating potential terrorists, and the way to protect ourselves from terrorism is to stop doing them." And I think this would get enormous support, if it were said not just in one speech but consistently and clearly, using all the resources available to the president. I suppose if the president said it, even Thomas Friedman would repeat it. After all, that's his job: to repeat what is said by a president he supports. I only pick Friedman because he's in many ways the most egregious of the surveillance supporters.

One thing that comes up periodically is frustration with the Democrats and the Republicans. People are exploring alternative parties. What are the pitfalls of going in that direction?

The first thing we should do is be realistic about the party system. Years ago, it used to be said, sardonically, that the United States has only one party: the business party, with two factions. That's no longer correct. It still has one party, the business party, but that party only has one faction. That faction is moderate Republicans. They're now called Democrats, but they're in fact what used to be moderate Republicans, since everything has shifted to the right. There is also another political organization, the Republicans, but they are barely making a pretense of being a normal parliamentary party. They're in lockstep service to wealth and power. They have to get votes somehow, so they've mobilized sectors of the population that they hope will be irrational and extremist.

The result is a population that is so confused and demoralized they just can't see what's in front of them. The most striking case of this is taxes. That's been polled for, I think, thirty-five years, and the results are consistent: a large majority favor higher taxes on the wealthy and corporations—the Democratic position.[10] Yet when asked to name the party they support on taxes, a majority say the Republicans. The same thing happens with security, health, and other issues. This is even true of so-called right-wing voters. Many of them support basic social-democratic policies, such as more spending on health and education, which is what the government does—but they don't support "government."

This confusion goes along with rising contempt for institutions—all institutions. Congress (single digits favorable), banks, corporations, science, anything. They're all against us. And some of the attitudes are really mind-boggling. Among people who call themselves Republicans, I think about half say that Obama is intent on imposing sharia law, not just on the United States but on the whole world, and about a quarter think maybe he's the Antichrist.[11]

Politicians are tapping elements of irrationality that are almost beyond description, including people who think we have to have guns to defend ourselves from the federal government. Rand Paul was recently trying to organize opposition to the U.N. small-arms treaty.[12] "Small arms" here means anything less than a jet plane. His opposition to the treaty is that it's a plot by the United Nations and the socialists, Obama and Hillary Clinton, to take away our guns so we won't be able to defend ourselves when the U.N. comes to take away our sovereignty. This is a guy who might be running for president. He's somewhere in outer space.

But that is what you find in a country that's become demoralized, confused, and overwhelmed by propaganda, from commercial advertising to national policy. Where the population is very much atomized, so people don't get together, don't interact in ways that are politically significant.

Solidarity.

Yes, solidarity. I don't want to exaggerate. There are plenty of people, including young people, who are committed to solidarity, mutual support, a unified struggle against the dangers we face.

I hesitate to call Occupy a movement, but let's use the term. Occupy has receded, clearly. Why do you think that happened?

I'm not so sure it has happened, frankly. I don't think it's clear at all. The Occupy tactic has receded. But that was obvious from day one.

The tent encampments.

You can't sustain that for long. You can for a while, but that's not the kind of tactic you can continue indefinitely. In fact, all tactics have a sort of half-life, and this one couldn't last more than a few months. But there's no question that Occupy lit a spark. There were hundreds, if not thousands, of such movements around the country, around the world, and they linked up with other such movements. And it's still going on. In early June 2013, during the Left Forum in New York, a parallel demonstration took place in Zuccotti Park in solidarity with simultaneous demonstrations in Greece, Spain, and Taksim Square in Turkey. That's solidarity. It's growing all over the world, with lots of mutual interaction and support. Much of the Occupy movement has gone into blocking foreclosure, neighborhood organizing, opposition to police brutality, fixing schools.

Also assistance to the victims of Hurricane Sandy in New York.

Yes, that made the papers. The Occupy organizers were the first responders, actually. We would like the movement to achieve a bigger scale, but Occupy hasn't gone away. These things are the hope for the future.

The Canadian poet and singer Leonard Cohen has a song called "Democracy," which says that it "is coming to the U.S.A." What will it take to make that happen?

The same thing it's taken for hundreds of years. Go back to the earliest democratic revolution in the modern period, in seventeenth-century England. In the 1640s there was a civil war, Parliament against the king. The printing press was available at the time. There were radical pamphleteers, itinerant preachers, radical movements such as the Levellers and others, and they were spreading their propaganda, their ideas, quite widely. The gentry, the ones who called themselves "the men of best quality," were appalled. They were appalled by pamphlets that said, "We don't want to be ruled by a king or Parliament, we don't want to be ruled by knights and gentlemen, who do but oppress us, but to be governed by countrymen like ourselves, who know the people's sores." And the gentry had to do something to stamp out democracy, which is always a threat.

Fast-forward a century to the American Revolution, so-called, and read the constitutional debates. James Madison and others were describing how to set up the constitutional system. The basic principle was enunciated by the president of the Continental Congress, John Jay, later the first chief justice of the Supreme Court. He said, "Those who own the country ought to govern it."[13] Or as Madison put it, power has to be in the hands of the wealth of the nation, the more responsible set of men who sympathize with property owners and understand that you have "to protect the minority of the opulent against the majority."[14] The rest of the population has to be tamed to make sure they can't do very much. That's the way the constitutional system was actually established, quite apart from slavery and the exclusion of women and so on.

Ever since then, there have been struggles about democracy. A lot has been gained, but every gain in freedom elicits a reaction from "the men of best quality." They don't give up power happily. They constantly find new ways to try to control and dominate. In the twentieth and twenty-first centuries, the main strategy has been to control opinion and attitudes. Huge industries, like the public relations industry, are devoted to this enterprise.

It's interesting to see how little recognition there is of some very obvious facts about the public relations industry. Its core activity is commercial advertising. What's commercial advertising? It's a way to undermine markets. Business doesn't want markets. Markets are supposed to be based on informed consumers making rational choices. That's the last thing businesses want. Take a look at a television ad, and it's completely obvious that it's trying to create an uninformed consumer, someone who will make a totally irrational choice—buy a Ford because some football player is standing next to it. The whole purpose is to undermine markets.

The same institutions run political campaigns and carry over the same ideas, techniques, and so-called creativity to try to undermine democracy, to make sure that you have uninformed voters making irrational choices. That's how you can get poll results like the ones I described.

When you compare attitudes and opinions with policy, you find a huge gulf. But much more interesting, you find a class basis for the divergence. Roughly the bottom 70 percent of the population in income level has no influence on policy. They're disenfranchised, so it doesn't matter what they think. Political leaders don't pay any attention to them. You go up the scale, people have more influence. You get to the real top, the one-tenth of 1 percent, they're basically designing the policies, so, of course, they get what they want. You can't call this democracy. It's plutocracy. It may be what Jim Hightower calls "radical kleptocracy." That's maybe a better term. Certainly not "democracy." The 70 percent doesn't need to read scholarly studies—they just know it doesn't matter what they think. "They don't pay any attention to us," they figure, which is correct.

You've written about a "democracy deficit."

"Deficit" is an understatement. Iran just had an election, and people criticized it, rightly, because you can't even enter the Iranian political system unless you're vetted by the clerics. That's terrible, of course. But what happens here? You can't enter the political system unless you're vetted by concentrations of private capital. If you can't raise hundreds of millions of dollars, you're out. Is that better?

2

A TOUR OF THE MIDDLE EAST

You were just in Lebanon. The dangers of a wider war in the Middle East seem to be increasing. The United States is now going to openly arm the so-called rebels in opposition to the regime of Bashar al-Assad. What did you learn on your trip?

Lebanon is quite interesting. People have somehow developed psychological defenses, so that they go on living extremely placid lives, as if they're not about to be consumed by a conflagration. But they are. Lebanon is a small country of four million people, and it has taken in more than half a million Syrian refugees. That's apart from the Palestinian refugees and Iraqi refugees already there. The country is under constant threat from Israel, which is quietly pointing out that it may decide to destroy all the missiles in Lebanon—Israel claims Lebanon has sixty thousand missiles scattered all over the country. What the Israelis say is, We learned the lesson of the last invasion.

We're not going to fight on the ground. The resistance is too strong.

The last Israeli invasion of Lebanon was in 2006.

Yes. This time, they say, We'll just get it done in two days. That can only mean bomb the country into rubble. But people act as if none of this is happening. Life goes on. Pleasant events, discussions.

On Syria, I'm really not convinced that our administration intended to arm the rebels in any serious way. If the United States and Israel wanted to support the rebels, there are very simple ways of doing it that don't involve sending arms. Simply have Israel mobilize forces on the Golan Heights—which is actually Syrian territory, though the U.S. government and the press call it part of Israel—forty miles from Damascus. In one day, you could march in. It's in artillery range.

Assad would be compelled to send forces to the south, as happened in the past when Israel mobilized forces. It would happen now. That would relieve pressure from the attacks on the rebels without sending one pistol across the border. Have you heard a word about it? It's not even discussed. It's not even an option. It's not that they can't figure it out. They can figure it out more easily than I can. But I think that means that they just don't want the Assad regime to collapse.

The United States and Israel are pretty happy, watching Arabs slaughter each other. It's deepening the Shiite-Sunni divide, which is tearing the region apart and is one of the worst consequences of the U.S. invasion of Iraq, a major crime. Let the Arabs kill each other and undermine each other. Meanwhile, we're around to pick up the spoils. And the Assad regime has been more or less in line with our interests. It cooperates in intelligence and has kept the border with Israel quiet. Maybe the U.S. and Israel don't love the situation, but I don't think they love the alternative either, which would probably be a jihadi-dominated government.

Incidentally, I did spend some time with Syrian democracy activists. Really wonderful people, impressive people, and very frustrated by the fact that they get almost no support from the West, including the Western Left, which doesn't support them the way it supported others in the region.

Why is that?

There are many reasons. One is that I think the activists are somewhat deluded about the situation in Syria. A lot of people on the left here think that the rebels are just trying to overthrow a legitimate government, maybe not the greatest government in the world but a legitimate government. Why should we support them? It's like the contras attacking the Sandinistas or something. That's a widespread attitude. You can argue about whether it's right or wrong, but it's certainly not without some elements of justification. But the rebels are not the same as the democracy activists I met. The activists think they are, but that's not the case.

It certainly started with them being the same. In the first couple of months of the uprising in Syria, it was a very impressive, honorable, popular movement, calling for reforms. It should have received support then, but it didn't. And soon it turned into a military confrontation. Once that happens, a

certain dynamic begins to develop: the harshest and most brutal elements come to the fore. They're the fighters. They know
how to kill, they're good at it. They come to the fore, and you
get increasing brutalization.

In the Vietnam War, for example, the National Liberation
Front were not saints but they were, in my view, the most hopeful and progressive element. Pretty soon, they were marginalized and ended up with essentially no power. In fact, I wrote
an article pointing out what seemed to be obvious at the time,
that the war would end either with the total destruction of
Indochina or else with the survival of only the most brutal elements, who would dominate all of Indochina.[1] Which is pretty
much what happened.

But that's what you can expect from a military conflict. And
I think we see it emerging in Syria, which is part of the reason
for the lack of support from the Left. It's maybe not a justification, but it's a reason. The young Syrian democracy activists—
the ones I met, at least—are in favor of the United States sending
arms to the rebels. They say that will partially equalize the military imbalance and drive Assad to negotiations, which will
then enable them to take over. But I think that's an illusion. First
of all, it's not going to equalize the military imbalance. As soon
as you send some arms to the rebels, more arms—and more
advanced arms—will come to the regime from Iran and Russia.
In fact, that's just what happened a couple of days after I was
there. There was an announcement that Iran had sent four thousand Revolutionary Guards and troops to support Assad.[2] So I
think sending arms to the rebels would raise the level of conflict, with the same imbalance.

The only faint hope that I can see—and it's pretty faint—is something like a Geneva negotiation, in which an agreement might be made between Russia and the United States to allow a transitional government in which the Assad regime participates and maintains some degree of authority, with the hope that the Assadists would be impelled to abandon political control and move toward some other system. The probability of this is really not high. But if there is a better alternative, I don't see it. And as far as I know, virtually every informed commentator sympathetic to the goals of the democracy activists says something like this, whether it's Patrick Cockburn, Robert Fisk, Jonathan Steele, Charles Glass, or others. I just don't see any other possibility. But it's not going anywhere, because, for one thing, the rebels say they would not take part in such a conference.

What about Israel? Viewed in the long term, the occupation seems self-destructive. And even former prime minister Ehud Olmert and former Shin Bet leaders have basically acknowledged this. So why does Israel persist?

I would question the word "acknowledged." The way they're presenting the situation, Olmert and the others, is that we either accept a two-state settlement or else there will be one state with the so-called demographic problem—too many Palestinians in a Jewish state. Either we will have to move to intolerable apartheid, they say, or else we will disappear. Those are the alternatives they offer.

The trouble is, those aren't the alternatives. It's a delusion. And I'm sure they know it. The alternatives are either a two-state solution along the lines of the international consensus—

or Israel and the United States continue doing exactly what they're doing right now.

And you can see that very clearly. The policy is explicit. It's being implemented before our eyes. First, separate Gaza from the West Bank. That's in violation of the Oslo Agreements, but who cares? It's a crucial step, because it means any autonomous government in the West Bank, however limited, will be cut off from the outside world. Gaza stays under a state of harsh siege, and as for the West Bank, Israel takes over the Jordan Valley—which is, in fact, what it's doing. Step by step, every couple of days, kick out another village, drill some more wells, and so on. Do it quietly, so the goyim don't notice—or at least pretend not to notice.

And then Israel will take over maybe 40 percent of the region that's left: the areas inside the so-called Separation Wall, which is actually an annexation wall; Greater Jerusalem, a hugely expanded area around Jerusalem proper; a couple of corridors extending through the occupied territories—one to the east of Greater Jerusalem through Ma'ale Adumim, which virtually bisects the West Bank, and one to the north that takes in the city of Ariel and cuts off most of the rest of the West Bank. Meanwhile, move the Palestinians out, but slowly, a village at a time, without fanfare or international publicity.

When all of this is integrated into Israel, there's not going to be any "demographic problem." There will be very few Arabs in the areas that Israel will ultimately integrate. No civil rights struggle, no anti-apartheid struggle. And the Palestinians will be left with a couple of small cantons that can supervise newspaper deliveries in the morning or maybe collect some taxes.

This has been going on for a hundred years. Quietly "create facts on the ground" without talking about it—that's been the traditional method of Zionist colonization. The Palestinians who remain are completely hemmed in. They don't even have access to Jordan, which is a U.S. client state.

There are some exceptions. In postcolonial systems, privileged elites have to be given a little piece of the action. If you go to the poorest, most repressed Third World countries, there's a privileged elite living in amazing luxury. That's what's happening in Ramallah, which is kind of like Paris and London. The Palestinian elite have a nice life there. So let that continue. That will kind of pacify them. And the rest of the population, let them rot.

That's the policy that's being carried out. That's the alternative to a two-state settlement. There is no one-state alternative. It's not an option.

Whatever Olmert may say, he's smart enough to know that Israel is not going to allow one state to emerge, for exactly the reasons he says. They don't have to, because they can continue the current policy. So, I hate to say it, but those who think they're helping the Palestinians by calling for one state are in practice supporting the continuation of the current policies, which may lead to some form of Palestinian autonomy, but of an utterly fragmented, meaningless kind.

Those are the alternatives—and that's what you have to face if you want to live in this world, not some world of abstractions in philosophy seminars.

Why does the U.S. government persist in its support of Israeli policies?

The primary reasons have been geostrategic. But Israel also has close links with U.S. military and intelligence. An illustration comes via WikiLeaks, which exposed a diplomatic cable listing sites of uniquely high significance to the United States. One was near Haifa: Rafael Advanced Defense Systems, manufacturer of drones and other high-tech military equipment. It's so closely linked to the U.S. military industry that it shifted its headquarters to Washington to be closer to the money.[3]

Israel has also been called upon to perform secondary services for the United States. For example, enabling Ronald Reagan to evade congressional restrictions to pursue his terrorist wars in Central America.[4] And it's highly valued by U.S. investors. Intel has established a major plant there for a new generation of chips. Warren Buffett recently said, after purchasing a major Israeli company, that "Israel is the leading, largest and most promising investment hub outside the United States."[5]

Apart from numerous such advantages, there are significant cultural factors. Elite Christian Zionism, based on biblical mythology, goes back long before Jewish Zionism. And particularly since 1948, it has been joined by the Zionist extremism of the vast evangelical movement, by now a substantial part of the Republican Party's base.

We should also not overlook another reason. The three countries that are most supportive of Israel are the United States, Australia, and Canada—all settler-colonial societies that virtually exterminated their indigenous populations. What Israel is doing seems quite consistent with their own national images.

Then there are the significant lobbies that support Israel: the American Israel Public Affairs Committee (AIPAC), the military industry, evangelicals, and others. In contrast, Palestinians lack all of these. They have neither wealth, nor power, nor support among the powerful, so they have no rights, by normal principles of statecraft.

Finally, one should not forget the dependency in the relationship, and its significance. When the United States puts its foot down, Israel must obey. That has happened repeatedly, from Reagan to George W. Bush.

What will it take to change U.S. policy?

The primary mechanism is the usual one: popular organization and activism. That can have an effect. There is also some concern in the military and intelligence about Israeli policies and their impact on U.S. interests. So far that concern has been mostly squashed, but it could become a factor. If the Arab oil producers or Europe were to pursue an independent course, that could also have an effect.

You were in Turkey in January. In late May 2013, street protests erupted, ostensibly over the building of a commercial mall at Gezi Park, near Taksim Square. Incidentally, part of the park was an Armenian cemetery that was seized by the government in 1939.[6] The mall project triggered deep-rooted resentments against the regime of Recep Tayyip Erdoğan. What do you see happening there?

I was in Istanbul to deliver the Hrant Dink Memorial Lecture. Hrant Dink, the Turkish-Armenian journalist, was assassinated, everyone assumes by the state, and there was a big backlash that led to a serious increase in concern about the Armenian massacre and its denial.[7] By now there is a pretty substantial popular movement interested in understanding and doing something about the Armenian genocide. And there was a huge demonstration in support of Hrant Dink and what he stood for, which the police didn't try to stop.

Taksim Square was already simmering at the time, and confrontation seemed imminent. The square is the last green, open area in Istanbul. The rest of the city has been hit by a wrecking ball of commercialization, gentrification, and authoritarian control that has essentially wiped out the commons. It's destroying an ancient treasure of ethnic neighborhoods and historical monuments, taking away public space in the interest of the rich.

Gezi Park is part of Taksim Square. When the bulldozers came in, there was resistance. People occupied the square, protesting against having the last piece of the commons destroyed. Erdoğan's reaction was like Hosni Mubarak's in Egypt or Bashar al-Assad's in Syria: send in the riot police and smash them to pieces. It was very violent.

Then Erdoğan kind of backed off a little, and for about a day it seemed as if a negotiated settlement was coming. Even the terms of it were announced: the government would wait for the outcome of a court case about the legitimacy of the demolition. If the court ruled that it was legal to proceed, there would then be a referendum in Istanbul.[8] That looked like a possible settlement. But within hours, Erdoğan sent the police in to smash everything up and drive the demonstrators out.[9]

Now there's a real split in Turkey between a conservative Islamic element, largely rural, and a secular, liberal, progressive element, which wants a more democratic and open society. The Erdoğan government has been becoming more repressive. Turkey has jailed more journalists than any other country.[10] You also see increasing Islamization, which many people don't like.

What happens in Turkey is of enormous significance. Turkey's importance in the region is substantial. And my own view is that there is a broader meaning to these developments. The human species at the moment is destroying its own commons. Nobody owns the atmosphere, for example. It's our common possession. The environment is a common possession—and we're destroying it.

It's a striking fact that the ones trying to defend the commons are mostly indigenous populations. They're in the forefront: the First Nations in Canada trying to block the tar sands, indigenous people in Bolivia and Ecuador, aboriginals in Australia, Adivasis in India, campesinos in southern Colombia. They're trying to protect the commons, protect the future for all of us. The richest and most powerful countries, like the United States and Canada, are happily destroying the commons.

What you see in Taksim Square is a microcosm of this. It's the same wrecking ball, just on a massive scale.

And the same thing is happening all over the world. Wherever you go, there's a battle under way between neoliberal depredation and the effort to protect future generations from this wrecking ball. The outcome is going to determine what happens to the species.

You say that if you look at the current political situation in the world, cynicism is justified but should not lead to passivity.

If cynicism leads to passivity, we walk off the cliff. That's what it means. The choices are stark: either you give up and help ensure that the worst happens or you become engaged and maybe you will make things better.

I don't know if you actually believe in reincarnation, but you have mentioned that if you could live in another era, you would like to be in Edinburgh during the Enlightenment. Why?

The Scottish Enlightenment was a period of unusual intellectual freedom, independence, thoughtfulness, reflection. It happened to be in Edinburgh mostly, and included thinkers such as David Hume, Adam Smith, Francis Hutcheson, others.

I wouldn't romanticize it too much. Hume is one of my favorite philosophers, but he wrote some essays that are pretty awful—like his essay on national character, which is very racist but rather favorable to Armenians. He says in their national character, Jews are known for "fraud" and Armenians for "probity."[11] Oh, I see why you brought up this question. You wanted me to say that.

So what about reincarnation?

I hope it isn't true. If there is reincarnation, what we should hope for is to be reincarnated either as bacteria or as beetles, because they're the ones who are likely to survive what we're creating in the world.

You end almost all of your talks with a few words about how people must organize and change never comes easily. But you don't go beyond that. When you give lectures, you talk for more than an hour, and then right at the end there's a coda of a few minutes, in which you say: "You can change things. Thank you." And it's over. People are sometimes left aghast.

Yes. "Tell us how to do it." Nobody can tell you how to do it. Nobody has ever been able to tell you. Saul Alinsky can give you some tricks for organizing a community, but that's not very much. Nothing you couldn't figure out for yourself in five minutes if you set yourself to it.

Furthermore, nobody from the outside can tell you what to do, because you're the one who knows the circumstances in which you live. You know what the options are. You know what can be done. You know who you are, what you're willing to undertake, how much commitment and engagement you're prepared to devote. Nobody can devote 100 percent of their time to political activism. So you're the one who has to decide. There's just no way out of that dilemma. You can't expect to find some savior coming from the outside and telling you, "Here's what you ought to do." That was true at every point in the past, and it's still true.

You travel so much, and when you're at home, you work incessantly. Do you reflect on your remarkable journey and the roads taken or not taken?

Not much. If I'm asked, I can think of things I should have done that I didn't do, but you go on. It's the normal way of life.

Do you ever feel like simply retiring and leaving all these political headaches to others to worry about?

It's going to happen pretty soon, whether I choose to or not.

3

POWER SYSTEMS
DO NOT GIVE GIFTS

CAMBRIDGE, MASSACHUSETTS (FEBRUARY 3, 2014)

We're speaking the day after the Super Bowl, the most watched event in the United States. A thirty-second ad costs $4 million. A huge bonanza for the media corporations—in this case, Fox.

I almost never watch television, but I did watch about ten minutes of the pregame show, where they have these zillions of ads. They're pretty interesting. They illustrate very well, actually, something that John Bellamy Foster and Robert McChesney wrote about: that as the economy moves toward oligopoly, greater concentration, more effort, is put into trying to prevent price wars, because they cut into profits.[1] So what you have to do is compensate by fraudulent product differentiation. That is, everyone produces the same products, but you have to sell them as if they're somehow different.

When you look at these ads, they are exercises in mass delusion, with enormous effort going into trying to get people to pick some commodity that they don't want, instead of this other commodity—which is identical to it—that they also don't want. Which is an interesting reflection of the way the society works.

There's a scene in the documentary Manufacturing Consent *in which you recall going to a sporting event and watching the reaction of your classmates and people in the crowd.*[2]

It's an interesting phenomenon, cheering for the home team. It's easy to get caught up in it, and it can be quite innocent. What's a little frightening, though, is the level to which people become dedicated to the victory of their own gladiators, people they have nothing to do with. When I was a kid, for example, the same players played with the New York Yankees every year, so there was a kind of a fraudulent but not totally ridiculous sense of identification with Joe DiMaggio or Lou Gehrig or such. But now a player can be on this team one year, a rival team the next year. You still have to cheer for your home team with enormous enthusiasm. If they lose, you descend into misery; if they win, you're exultant. Though it can be innocent pleasure—that's not impossible—it can also be pretty dangerous, fostering blind allegiance.

You had an interesting experience you told me about, when you were in the fourth grade. You went with a certain teacher to see the New York Yankees and the Philadelphia Athletics.

Miss Clark. Every boy in the fourth grade was in love with Miss Clark. She took me and my best friend to a baseball game, which was an unheard-of pleasure. If you want to be bored, I'll give you an inning-by-inning account of it. We sat in the cheapest seats, in the bleachers right behind Joe DiMaggio. Of course, we wanted the Yankees to lose, because we were from Philadelphia, but nevertheless there were all these heroes out there—Lou Gehrig, Bill Dickey, Red Ruffing. The A's weren't up to that level, but they had a couple of quasi-heroes. So we were just ecstatic. Except a couple of months later, she betrayed us. She married the art teacher, Mr. Fink. I never got over that.

There was a notable outcome to the game, as well.

We were ahead 7 to 3 until the seventh inning, when the Yankees scored seven runs and won, 10 to 7. Boys of my age who lived in Philadelphia had a kind of an inferiority complex because the Philadelphia teams lost in every sport. But to make it worse, our cousins were all in New York, and they were at the top in every sport. So we had to survive this interaction with these cousins who were lording it over us that they won everything and we lost everything.

In the third grade you had an incident where you copied something from the Encyclopaedia Britannica. *Do you remember the details of that?*

How do you know all these terrible things about me? Yes, that was one of my real crimes. We had an assignment to write something about astronomy. And I don't know why, but I copied a section out of the *Britannica* and handed it in. I didn't think anything about it at the time, but later I felt very bad. I was never censured. The teacher must have known I couldn't have written it, but I never heard anything about it. And I've been trying to live that down all my life. It's almost as bad as the A's being defeated by the Yankees or Miss Clark betraying us.

And then to go back to what might be your first act of rebellion, refusing to eat oatmeal. How old were you and what were the circumstances?

That I can date, because I know where it was. I was a year and a half old. My relatives were mostly working-class, unemployed, and my parents were teachers, which meant they had an income. So the relatives tended to congregate around our house, especially over the summer. One of my aunts was trying to feed me oatmeal. I was sitting on a counter, and she was forcing oatmeal into my mouth, and I didn't want to eat it. So I put it in my cheek and I just kept it there and refused to swallow it. I don't know how long that went on, but I can remember trying really hard not to swallow that oatmeal.

And you haven't looked back since.

It's still there.

From a very early age, you were attracted to anarchism. What is it about anarchism that appealed to you?

Anarchism seems to me self-evident. Why should structures of authority exist? Every structure of authority, hierarchy, or domination has the burden of proof; it has to demonstrate that it's legitimate. Maybe it can. If it can't, it should be dismantled. That seems to me about as close to a truism as I can imagine. And that's the dominant theme of anarchism: identify structures of power and domination, whether a patriarchal family or an imperial system or anything in between, and demand that they justify themselves. And when they cannot, which is almost always, move to dismantle them in favor of a more free, cooperative, and participatory system. This just seems intuitively obvious.

And you discovered anarchism while rummaging through bookstores in New York?

By the time I was about eleven or twelve, my parents would let me go to New York by myself. I would take the train in and stay with relatives. I'd go on weekends or whenever I was off from school. In those days—it's very different now—Union Square was kind of grubby and, among other things, there were little anarchist offices there. *Freie Arbeiter Stimme*, the Yiddish anarchist newspaper, had an office there. And I'd hang around. They'd have pamphlets. People were happy to talk to you.

And then right below Union Square, down on Fourth Avenue, which was also pretty grubby then, there were rows of little stores, including secondhand bookstores run by émigrés. Among them were refugees from the Spanish Civil War, anarchists who fled after the revolution was crushed, in 1936. To me they looked about a hundred years old. They were probably thirty. They'd had interesting experiences. They were eager to talk. And they also had a lot of pamphlets and other literature. I didn't have much money, but the stuff was cheap enough to buy, and I collected a lot of it.

Remember, this is the 1930s and the early 1940s, a very lively period of radical journalism and radical discussion. Actually, the downtown Philadelphia public library had a really good selection of current radical publications. I'd sometimes go down there on a Saturday afternoon and rummage through their collections.

One of the anarchist thinkers who influenced you was Rudolf Rocker. He was born in 1873 in Germany, and he died in upstate New York in 1958. Did you ever meet him?

I never met him. When I was a kid, back in those secondhand bookstore days, I did find a couple of pamphlets by Rocker. But I only found his *Anarcho-Syndicalism* somewhat later, probably in the late 1940s or maybe even 1950.[3] It was actually written in 1938, but I don't think it was really available till maybe ten years later. I thought it was a very insightful work.

Rocker wrote, "Political rights do not originate in parliaments, they are, rather, forced upon parliaments from without."[4]

From below, in fact. I think that's an accurate comment. Power systems do not give gifts willingly. In history, you will occasionally find a benevolent dictator, or a slave owner who decides to free his slaves, but these are basically statistical errors. Typically, systems of power will try to consolidate, sustain, and expand their power. That's true of parliaments, too. It's popular activism that compels change.

In "Notes on Anarchism," which you wrote in the early 1970s, you say, quoting Rocker, that " 'freeing man from the curse of economic exploitation and political and social enslavement' remains the problem of our time."[5]

Very much so—and to this day. We can add to that an observation that's typical of the anarchist tradition, and was made by Karl Marx, that overcoming the animal problems of survival, exploitation, oppression, and so on will free us to be able to face our human problems.

In another essay, "Language and Freedom," you wrote that capitalism "is not a fit system. . . . It is incapable of meeting human needs."[6] *What is it about capitalism that allows it to keep going forward? What sustains it?*

What sustains it are two tendencies. First, the inclination of those with enormous power to secure and maximize their power. That's one tendency. The other is the passivity, hopelessness, or atomization of the people below—the ones Rocker wrote about—who could force a change. I wrote that essay in 1970, which happened to be the beginning of a major backlash against the liberatory character of the 1960s. That huge backlash, which we're still in the middle of, was the beginning of the neoliberal assault on the population of the world.

There were things I didn't know at the time, things that everyone ought to know now—namely, that we are facing a severe environmental crisis. Every issue of a science journal that you read has more alarming discoveries about the threat confronting us and the imminence of it. It's not hundreds of years away; it's decades, maybe. And yet predatory capitalism is telling us to maximize the threat, to extract every drop of fossil fuel out of the ground.

The excuse is jobs. But in modern political discourse, the word "jobs" replaces an unspeakable, obscene seven-letter word: "p-r-o-f-i-t-s." You can't say that, so you say "jobs." We have to make sure we get jobs. Because the power systems care so much about working people, you see. That's why we have to race off the cliff like the proverbial lemmings.

Major sectors of the corporate system—the Chamber of Commerce, energy corporations, and so on—openly announce that they are carrying out massive propaganda efforts to try to convince people that there is no climate change or that, if climate change does exist, it's not anthropogenic. It's not because of humans, it's because of sunspots or something.

These efforts to drive people toward complete irrationality and self-destruction are enormous and growing. Some of them are almost surreal. One example is ALEC, the American Legislative Exchange Council, a corporate-backed group that writes legislation for states. They figure it's easier to coerce the states than the federal government, so they try to force extremely reactionary state legislation "in favor of jobs"—namely, that unspeakable word.

One of their programs is to "provide instruction in critical thinking" in grade schools.[7] Who can be against that? Well, how do you bring critical thinking to the schools? If there's a sixth-grade class that has something about climate change, they say, you should also introduce into the curriculum something against the theory of climate change, so the sixth graders will learn to think critically, to evaluate the opinion of 99 percent of scientists on one hand versus half a dozen skeptics and the major corporations on the other. That will teach "critical thinking."

Let me just add, ALEC is largely funded by the billionaire Koch brothers.[8]

The efforts that go into trying to ensure the end of humanity are impressive. If there were somebody from outer space watching this, they could only conclude that humans are an absolutely unviable species, an evolutionary error tending toward self-destruction.

Is it even possible to protect the environment under capitalism?

Our economic system has deep institutional properties that drive toward destruction. That's even part of economic theory. In a market system, you pay no attention to what are called "externalities." If you and I make a transaction, let's say you sell me something, we each try to maximize our own benefit. That's how the system is supposed to work. We don't ask about the effect of the sale on other people.

Take Goldman Sachs. When they make a risky transaction, they probably try to cover themselves for the risk, but they don't pay attention to *systemic* risk—that is, the threat that if their transaction goes bad, the whole system will collapse, like what happened with AIG, for example.[9] In a way, they don't have to worry about that. If necessary, the government is going to bail them out, so it's fine. That means that risk is underestimated, because externalities are ignored. And that can be devastating. In fact, that was part of the reason for the global financial crisis.

The Koch brothers, or even those less extreme than they are, are driven by the desire to make a profit. That's the nature of the system. If you're a CEO or on a board of directors, you're supposed to make a profit. You don't pay attention to the costs to others. And in the case of the environmental crisis, one of these costs may be destroying our species. It's an externality, so therefore it's a footnote. Of course, when it comes to the environment, there's nobody to run to, cap in hand, to ask for a bailout. In a financial crisis, the taxpayer can be bamboozled into bailing you out, but not in the environmental crisis.

Can it change? Sure. The economic system is not a law of nature.

Given the severity and the urgency of the environmental crises, where do you see the clamor for change?

There is clamor. There are protests at the White House, there's a lot of local resistance to the huge expansion of pipeline networks in many parts of the country.[10] But so far none of the resistance is at a scale that can compete with the vast economic resources and influence of the major energy corporations. That's why the newspapers present the issue of climate change as a kind of a he says, she says issue. Maybe it's happening, maybe it isn't. It's true, you can never have absolute certainty in the sciences, but this is about as close to overwhelming consensus as anything imaginable.

I think the Intergovernmental Panel on Climate Change used the figure of 95 percent certainty.[11]

Yes. And the general agreement with it is overwhelming. There are a few critics—they get plenty of publicity—who raise questions about, say, methodology. But there is a much more significant group of critics who are almost never mentioned. They're the ones who think the IPCC reports are too conservative. Remember, "uncertainty" means it might not be as bad as predicted—or it might be worse. The way it's presented, "uncertainty" about climate change is taken to mean that, well, maybe things won't be as bad as expected. But if you go to many climate scientists, Michael Mann at the Earth System Science Center at Pennsylvania State University and others, they think the IPCC estimate is too rosy.[12]

Some of the things that can be done to counter climate change are elementary. Take weatherization—making energy-efficient homes. Not only would it delay the environmental crisis, it would help overcome the employment crisis. There are tens of millions of people who can't get work. Their lives are being destroyed, their children's lives are being destroyed. And they could be put to work weatherizing houses.

There are innumerable opportunities like that. But almost nothing is being done. You hear great enthusiasm about the United States becoming the Saudi Arabia of the twenty-first century. We'll have huge energy resources, people say. But what is that going to lead to? We'll get cheap energy as we race toward self-destruction. And we'll take other countries down with us. The Europeans, who have been trying to do something significant about climate change, are backing off from those efforts because they can't compete with the low energy prices in the United States.

And it's not just Europe. Take Ecuador, a poor Third World country. It has a fair amount of oil located in ecologically threatened areas. Ecuador asked the rich countries to provide just a few billion dollars, a small percentage of what the oil would sell for, to enable them to keep it in the ground, which is where it belongs. But they couldn't get the money. The rich countries won't do it.[13]

This is the fiftieth anniversary of Lyndon B. Johnson declaring the War on Poverty. Why are there still so many poor people in the United States?

The War on Poverty was successful, and did cut back poverty notably. But then as soon as the neoliberal onslaught started, the trends reversed. Incidentally, that's not just in the United States. The whole world was subjected to neoliberalism in different forms. Europe is even more extreme than the United States right now. Imposing austerity during a recession, which even the International Monetary Fund says is economically unacceptable, has the very specific effect of dismantling the welfare state, Europe's great achievement in the postwar period. The business world and the wealthy don't like the welfare state—and never have. The call for austerity gives them an opportunity to get rid of it.

Why the attacks on food stamps and cutting unemployment benefits? Why this seeming sadism?

Not just that. Why the attack on Social Security? Why the attack on public schools? Anything that might benefit the general population has to be cut, because the goal of society must be to enrich and empower the rich and powerful, period. It all has a common theme. What you call sadism is something a little different, I think. It's trying to undermine the heretical, subversive conception that you ought to care about other people. You've got to get rid of that notion. You should just care about yourself or the powerful whom you're serving. Therefore, schools, Social Security, food stamps, all of these things are really subversive.

Why should I pay taxes for schools? I don't have kids in school; why should I pay taxes just so the kid across the street can go to school? This is what's called libertarianism in the United States. That's their doctrine. Why should I pay for something that I don't personally benefit from? Why should there be Social Security? Why should there be food stamps? Those guys should get out and work like I do, even though I'm getting all my profits from the taxpayer via the financial system.

These doctrines all are of a piece. It's kind of striking that they're exactly the opposite of what their heroes believed. Adam Smith and the founders of classical liberalism took it for granted that the fundamental human drive was sympathy and mutual support. What's called libertarianism today is the exact opposite: the drive to get rid of those subversive ideas.

This year marks the thirtieth anniversary of our very first interview. The topic was politics and language. I guess you must hold the world record for being interviewed.

I don't know what the world record is, but it's impossible to keep up. Every night I get a dozen requests.

And when you do all these interviews, is there anything you specifically want to inject into the conversation?

The requests for interviews come from a wide variety of sources. And what I try to do, however successfully—that's for others to judge—is to see if I can identify assumptions that are taken for granted but that are questionable and ought to be challenged and interrogated.

You once said that Amos was your favorite prophet. What is it about Amos that attracts you?

The word "prophet," first of all, means pretty much what we mean by "intellectual." It doesn't have anything do with prophecy. Amos opens by saying, I am not a prophet and my father is not a prophet. I am a simple shepherd and farmer.[14] Then he goes on to say some very profound things. I liked that.

Is the word in Hebrew navi?

Navi. It's translated as "prophet," but that's a very dubious translation of an obscure word. Nobody really knows what it means. These people didn't prophesize. They did geopolitical analysis, condemned the evil kings, the power structures, and called for care for the oppressed and mercy for widows and orphans. They were what we today would call dissident intellectuals. And they were treated like dissident intellectuals. They were driven into the desert, imprisoned, condemned.

One of my favorites is Elijah. He's the original person guilty of Jewish self-hatred. King Ahab asked him—the right translation would be "Why are you a hater of Israel?" "A hater of Israel" means he condemned the king. That's the origin of concepts such as being anti-American, anti-Soviet. If you are deeply totalitarian, you identify the rulers with the society, the culture, and the people. So if you condemn the rulers, you're against the society.

You have also been called a "self-hating Jew," of course.

I'm happy to be associated with Elijah, who was opposed to the most evil king in the Bible.

In 1953, you and your wife, Carol, were living on a kibbutz in Israel. For a while you were considering moving to that country. What happened to change your mind?

We were there just for a couple of months in 1953. We were students, so we took off for the summer. Yes, we were thinking about it. I had just been appointed here at MIT, so Carol went back by herself and stayed for a longer period. She came back to the United States, assuming that we would go back for good. We thought about it, but we just didn't do it. It wouldn't have worked for very long, I don't think.

We were in a very left-wing kibbutz that was sort of a center of outreach to the Arab community. There were a lot of things I liked about it, other things I didn't. But over the years, it changed a lot. It's now very reactionary. I couldn't possibly have stayed.

And I remember you telling me there were traces of racism.

They were pretty strong, yes. Just to give you an illustration: there was a group of young Moroccan Jews, who, I later discovered, were pretty much kidnapped from their parents. We lived among them. The kibbutz was very poor then. We lived in a packing container. And we were warned by others on the kibbutz, "You've got to lock your door and watch out for the Moroccans. They're a bunch of criminals." They were perfectly nice kids.

Once I was working out in the fields, picking grapes or something, and there was an altercation among some teenagers. The woman who was in charge of the agricultural section walked over to find out what was going on. When she came back, I asked her what had happened. She said that the kids from the kibbutz were bullying these other kids, who they thought were Moroccan Jews. "But I had to tell them, they weren't Moroccan Jews, they were Arabs who were visiting, and we had to be nice to them." If they had been the Moroccan kids, it would have been all right.

There were a lot of things like that. I once went along with someone who was going around to Arab villages, to get votes for his political party. He was kind of like a ward heeler. Back then, I understood enough spoken Arabic to be able to follow the conversations. In one village, I remember, the people lived across the road from a kibbutz, which they were sort of friendly with. They wanted to have commercial and other interactions with this kibbutz, but they couldn't cross the road unless they went to Haifa—which was maybe twenty miles away—and got a permit to cross the road. Then maybe they would be allowed to go across and interact with the people on the other side.

Some kibbutz members objected to that, but most didn't.

Another time, I was working out in the fields with an older member of the kibbutz, and I noticed a pile of stones. When I asked him about the rubble, at first he didn't say anything. Later, he took me aside and explained that this was actually the site of a friendly Arab village that had been destroyed in 1948. He felt guilty about destroying the village, but he said, "There were Arab tanks a couple of miles away. We just didn't want to take any chances."

You've commented to me that the Israeli Left is "almost nonexistent." Explain that. In some quarters—in the United States, at least—there was a perception for years that there was a very vibrant Left in Israel. What's happened?

There was a vibrant Left. It's seriously declined. There are a few very good people—honorable, courageous people—who remain, but the Left is scattered. So one of the best, for example, Amira Hass, a wonderful journalist, lives in Ramallah. She doesn't want to live in Israel anymore. Quite a few close friends of mine who were really committed to Israel have left. They were born there, they wanted to stay. But they couldn't stand it anymore.

The country has moved very far to the right. In fact, what's happening in Israel is quite similar to what happened in South Africa. You can read history from about, say, 1960 to the present and replace "South Africa" with "Israel," and it almost describes what's happening.

Back around 1960, we know from declassified documents, the South African foreign minister called in the U.S. ambassador and told him, Look, we know apartheid is being condemned around much of the world. We're becoming a pariah country. Of course we're right, but they don't understand us. We're being voted down in the United Nations. But it really doesn't matter, because there's only one vote that counts: your vote. You and I know that. As long as you support us, it doesn't matter what the rest of the world thinks.[15]

And that was true. In the 1960s, there was an anti-apartheid movement in England. But the United States kept supporting the apartheid regime. Reagan in particular, who was an extreme racist, refused to believe that there was any race issue in South Africa. His view was that it was just tribal warfare: the Zulus, the whites, and others were engaged in some kind of tribal conflict.

The South Africans were trying to intimidate and impose their rule on the surrounding countries, to create supportive client regimes. They were carrying out murders and aggression in Angola, illegally occupying Namibia, committing atrocities in Mozambique. And Reagan and Margaret Thatcher (though she was less of a fanatic than he was) were supporters of apartheid right to the end. In fact, Reagan was supporting terrorist groups in Angola—UNITA was essentially a terrorist gang—even after the South Africans pulled out.

By the 1980s, the apartheid regime was becoming delegitimized. There were boycotts and sanctions. Even the U.S. Congress passed sanctions, which Reagan had to veto.[16] I'm sure you know that the ANC, the African National Congress, was condemned by the United States in 1988—almost at the end of apartheid—as one of the "more notorious" terrorist groups in the world. That was shortly before Mandela was released. In fact, Mandela himself stayed on the terrorist list until 2008. It took special legislation in Congress to get him off it.[17]

But ultimately the apartheid regime and their supporters couldn't get away with it. And the reason was Cuba—though that's something we don't talk about in the United States. If you remember, when Nelson Mandela was released from prison, almost his first words were to praise Fidel Castro and the Cubans for being an inspiration and to thank them for playing an enormous role in liberating Africa and ending apartheid.[18] The Cubans drove the South Africans out of Angola and compelled them to leave Namibia.[19] Mandela pointed out that this had a very important psychological effect, in both black and white Africa, because it destroyed the myth of the invincible white man.

If you think this through, it's pretty similar to what's been going on in Israel, though it's happening a little later there. By the early 1970s, Israel had to make an important decision. In 1971, in fact, it was offered a full peace treaty by Egypt.

By Anwar Sadat.

Basically, a full peace treaty, normalization, everything, if Israel would pull out of the Egyptian Sinai. Israel considered the proposal and rejected it.[20] They had big plans to settle and develop the Egyptian Sinai at the time, drive out the Bedouin population, and build all-Jewish cities and kibbutzim. They decided to favor expansion over security. A peace treaty with Egypt—the only military force in the Arab world—would have basically meant complete security. That was a fateful decision.

Since then—we can run through the history—it's been the same. The more you pursue expansion and reject diplomacy, the more you're going to become isolated. And the last resort is the United States. Israel is essentially taking the position that South Africa did. We don't care about the world. We're right, the world is wrong, they're all anti-Semites. It doesn't matter if they're all against us, as long as you support us.

You've commented that people who say they are supporters of Israel are actually contributing to its destruction.

I've been saying that since the 1970s, when they made these decisions to pursue expansion rather than security and diplomacy. Now they're deeply concerned about what they call "delegitimization." Benjamin Netanyahu harshly condemned John Kerry because he referred to the fact that Europeans are beginning to boycott activities involving the illegal settlements.[21] Even to *refer* to that fact is anti-Semitic. That's pretty similar to what you heard from racist South Africans as the whole world closed in on them.

There are also movements led by Code Pink in this country against Israeli cosmetic products that are made in the occupied West Bank.

Yes, there are some activities here, though much more is happening in Europe. One of the major Danish banks canceled its deals with Bank Hapoalim because of its activities in the settlements.[22] The European Union has passed resolutions—I don't know if they're going to implement them—refusing contact with any Israeli institutions involved in the settlements.[23] And the "delegitimization" is expanding. Here to a certain extent, too.

Those who are pro-Israel, for lack of a better phrase, say that the country is being unfairly singled out, that there are double standards, and the like. Do you give any credence to those views?

That's what supporters of apartheid said, too. Why condemn apartheid? Look how awful China is. In fact, if you went to the old Soviet Union, you found similar criticisms of dissidents. Why are you condemning what we do in Czechoslovakia? Look at what the United States does in Central America, which is much worse. That's a standard position of those who support atrocities.

There's a perfectly obvious reason to scrutinize Israel. Yes, there are terrible things going on in other places, but in Israel we're helping carry them out.

In what way?

First of all, there's the $3 billion in aid each year, which is really probably about twice that figure when you examine the details of how it works. Then there is our diplomatic support: the vetoes at the Security Council to protect Israel, pretty much like Reagan's vetoes of Security Council resolutions condemning South Africa. We also have very close, intimate military relations, much more so than we did with South Africa. The Israeli government can get away with what they're doing only because the U.S. government supports it.

So the reason for focusing on Israel is that one should focus on oneself. In Iran, you expect dissidents to talk about Iranian crimes, not Israeli crimes. In China, you expect Ai Weiwei to talk about China and not the Congo. I suppose the commissars everywhere say, "Ah, double standards." But the reasoning is the same. We respect dissidents in other countries who focus on their own crimes. It's just ourselves we're not allowed to examine.

What do you think of the Boycott, Divestment, Sanctions movement? Do you support it?

In the case of South Africa, it's interesting that there wasn't a BDS movement. There were BDS tactics used, but there was no BDS movement. The tactics were, by and large, pretty successful and were targeted. So, for example, South African academic institutions were targeted for things like racist hiring. Sports teams were targeted because they wouldn't allow blacks to participate.[24] In general, products were boycotted if their manufacture involved apartheid conditions. There were condemnations of the way black workers were treated, of the Bantustans, and so on. That was all pretty effective. The United Nations banned weapons shipments to South Africa.[25] That's a boycott. But there was no BDS movement.

Here—and it's mostly the United States—there is a movement that is very ambiguous. It does not distinguish the kinds of things that are principled and effective from feel-good activities that are ineffective and even harmful. For example, boycotting products from the settlements or research activities conducted together with the settlements—anything connected with the illegal settlements—makes perfect sense. It's intelligible, it brings out the issues, it strikes at the crucial point, the illegal occupation, and it's effective. The kind of delegitimization that Israel is worried about now is protests about the settlements.

On the other hand, if you start protesting, say, Israeli discrimination within Israel, which there is, then it's kind of meaningless. How about within the United States? Why don't we boycott Harvard because of the number of blacks in prison? Apart from being empty and meaningless, it's also counterproductive.

The American Studies Association voted to boycott Israeli academic institutions.[26]

And the immediate reaction, perfectly predictable, was a huge backlash: "It's all anti-Semitism. We have to support Israel."

So this was no help at all to the Palestinians. When you're an activist, you have to think about the people you're trying to protect, not just make yourself feel good. That's elementary. Not every action that makes you feel good is going to be helpful to the victims. Some might even be harmful. Once you get a BDS movement, you're going to be vulnerable to that kind of problem, because movements have leaders that you have to follow, have a catechism that you have to repeat, and so on. It tends to be a very mixed story. Some of the tactics can be effective and sensible, and important and helpful to the victims; some can be harmful. And you have to distinguish between them.

Take, say, the Vietnam protests. You could understand why young people were so outraged and desperate by the late 1960s that they decided the way to protest the war was to walk down Main Street and break windows. The Vietnamese were strongly opposed to that. They understood that going down Main Street and breaking windows was just going to create a backlash in support of the war.

In those days, the Vietnamese advocated tactics so mild that the U.S. movement laughed at them. I remember meetings where the Vietnamese talked about how impressed they were when a group of women stood at a graveyard and mourned the U.S. soldiers who had died. It was pretty hard to sell things like that to the activists here, but the Vietnamese didn't care whether American activists felt good. They wanted the war to end. If you can't think about that, then don't call yourself a committed activist. These are the things you have to keep in mind. What are the effects going to be?

A listener sent me this Howard Zinn quote: "Small acts, when multiplied by millions of people, can transform the world."[27]

Yes, that was one of the main themes of his work. There are plenty of examples. In 1960, a couple of black students sat in at a lunch counter in Greensboro, North Carolina. They were, of course, immediately arrested and thrown out, which could have ended the protest. But the next day, more black students came in, which led to more arrests. And then came the Freedom Riders and the Student Nonviolent Coordinating Committee (SNCC). Pretty soon you had a massive civil rights movement. The courageous actions of the students in Greensboro were one significant contributing factor. The sit-in didn't end racism, but it achieved a lot.

What are some techniques for breaking through the barrier of received wisdom?

The first step is developing an open and critical mind, taking the doctrines that are standard and questioning them. Is the United States dedicated to democracy? Is Iran the greatest threat to world peace? Do we have a market system? Does the public relations industry try to promote choices or to restrict them? Anything you look at, every one of these things, you have to ask yourself: Is it true? A pretty good criterion is that if some doctrine is widely accepted without qualification, it's probably flawed.

So once you've taken the first step, and are willing to challenge dogma, then you can start reading more and looking at the world with more informed and open eyes. And then you have to join with others. There's not a lot you can do yourself. Going back to Howard Zinn's comment, joining with others to carry out those small actions that can light a spark has been very successful in the past. There is no reason it shouldn't be in the future.

To return to power systems, whenever they come under any kind of pressure, they offer up reforms. These, to a great extent, are placebos or hot air. You told me years ago that when you hear the word "reform," you should reach for your wallet because someone is probably trying to steal it.

"Reform" is an interesting word. Like with most terms of political usage, you have to distinguish between its literal meaning and its meaning in political warfare. "Reform" is usually used to mean something that power systems approve of. Changes they disapprove of aren't called reforms. So, we don't call Mao's collectivization programs "reforms." On the other hand, you can read praise of Mexico's "reforms"—namely, opening the oil industry to international exploitation instead of maintaining it for Mexico.[28] And "educational reform" refers to the various measures that are being undertaken to undermine the public education system in the United States. So, yes, you have to be careful about the term.

On the other hand, this shouldn't lead us to overlook the fact that changes imposed on systems of power by public pressure sometimes do ameliorate conditions and are literally reforms. This can happen under conservative as well as liberal administrations. Take Richard Nixon. Under Nixon, legislation passed that was quite effective: the Environmental Protection Agency was established, and so was the Occupational Health and Safety Administration. The earned income tax credit was put into motion, which is maybe one of the most important welfare programs. These reforms did not happen because Nixon was a nice guy. They came about under pressure. But they're significant. They weren't revolutionary, they didn't change the institutional structures, but they modified them and made people's lives better.

Let's change subjects. I do crossword puzzles. I wonder if you ever do them.

No.

Do you know that you are sometimes a clue in the New York Times *crossword puzzle? It's always "linguist Chomsky," and it's a four-letter answer.*

I wonder what it is.

But it's interesting that it's never "dissident Chomsky" or "social critic Chomsky."

I know about that because Carol used to do the puzzles. I think they're a horrible waste of time.

Sometimes you learn things. I just came across, in a recent puzzle—this is the clue, so see if you get it—"Philosopher who wrote 'It is difficult to free fools from the chains they revere.' "

I can think of many people who might have said that.

It was Voltaire.

That's interesting. But you can learn things much more easily just by opening the pages of a serious book.

4

ISIS, THE KURDS, AND TURKEY

The Middle East is engulfed in flames, from Libya to Iraq. There are new jihadi groups springing up all the time. The current focus is on Islamic State, or ISIS. Can you talk about ISIS and its origins?

There's an interesting interview with Graham Fuller, a former Central Intelligence Agency officer, one of the leading intelligence and mainstream analysts of the Middle East.[1] It's titled, basically, "The United States Created ISIS." Fuller hastens to point out that he doesn't mean the United States deliberately decided to bring ISIS into existence and funded it. His point—and I think it's accurate—is that the United States created the conditions out of which ISIS developed.

Part of it was just the standard sledgehammer approach: smash up what you don't like. In 2003, the United States and Britain invaded Iraq, a major crime. Iraq had already been virtually destroyed, first of all by a decade of war with Iran—in which, incidentally, Iraq was backed by the United States—and then by a decade of sanctions. The sanctions were described as "genocidal" by two respected international diplomats who administered them, and who resigned in protest for that reason.[2] They demolished civilian society, strengthened the dictator, and compelled the population to rely on him for survival. Finally, in 2003, the United States decided to just attack the country outright, an attack compared by many Iraqis to the Mongol invasion of nearly a thousand years earlier. Hundreds of thousands of people were killed, millions made refugees, millions of others displaced, the archaeological richness and wealth of the country destroyed.

One of the effects of the invasion was to institute sectarian divisions. If you look at a map of Baghdad in, say, 2002, it was a mixed city: Sunnis and Shiites lived in the same neighborhoods and intermarried. In fact, Iraqis sometimes didn't even know who was Sunni and who was Shiite. It was like knowing whether your friends are in one Protestant denomination or another. There were differences, but no hostility. In fact, both sides were saying, "There will never be Sunni-Shiite conflicts. We're too intermingled in how we live, where we live, and so on." By 2006, however, there was a raging sectarian war in the whole region, with Sunnis, Shiites, and Kurds separated from one another and set at each other's throats.

The natural dynamic of a conflict such as this is that the most extreme elements begin to take over. Their roots are in Saudi Arabia, a major U.S. ally and the most extremist, radical Islamic state in the world. Saudi Arabia makes Iran look like a tolerant, modern country by comparison. Not only is it ruled by an extremist version of Islam, the Wahhabi/Salafi version, but it's also a missionary state. It uses its huge oil resources to promulgate these doctrines throughout the region, funding clerics and setting up schools and mosques from Pakistan to North Africa.

ISIS comes ideologically out of the most extremist form of Islam, the Saudi version, and it's also funded by Saudi Arabia— not the Saudi government but wealthy Saudis, along with wealthy Kuwaitis and others, who provide the money and the ideological support for the jihadi groups that are springing up all over the place. But politically, it comes out of the conflicts engendered by the U.S. smashing up Iraq, which have now spread everywhere. That's what Fuller meant by saying the United States created ISIS.

You can be pretty confident that as conflicts develop, they will become more extremist. If the United States manages to destroy ISIS, we will have something even more extreme on our hands.

In Manufacturing Consent, *you observe: "A propaganda system will consistently portray people abused in enemy states as worthy victims, whereas those treated with equal or greater severity by its own government or clients will be* unworthy."[3] *Then you give the example of the Kurds in Iraq and the Kurds in Turkey.*

The Kurds in Iraq first became victims of U.S. power in the 1970s, when the United States essentially sold them out to Saddam Hussein. In 1974, as a favor to Iran, Washington supported a Kurdish rebellion against Iraq. But a year later, Iraq and Iran made a deal, and the United States just stepped back, leaving Iraq free to massacre the Kurds. When Henry Kissinger was asked why we did that, he made his famous statement "Covert action should not be confused with missionary work."[4]

Through the 1980s, Saddam was a major U.S. ally, and the United States supported him in the war against Iran. He was taken off the terrorist list in 1982 so that the U.S. could start providing him with aid. As is well known, he then launched a horrendous attack against the Iraqi Kurds. The Reagan administration, including Reagan himself, blocked efforts even to criticize the attack. The Pentagon came out with a story that it was Iran that was responsible for the Halafbja massacre, the al-Anfal campaign, and the other atrocities.[5]

Support for Saddam continued under George Bush no. 1, the one who is called the statesman, George H. W. Bush—not the madman, George W. The first Bush just adored Saddam Hussein. He overruled Treasury Department objections to sending more agricultural aid to Saddam, badly needed in part because Saddam had devastated large Kurdish agricultural areas.[6]

In August 1990, Saddam made his first mistake. He disobeyed U.S. orders—or, more likely, just misunderstood them—and invaded Kuwait. The reaction was very strong. He immediately recognized his mistake and tried to find a way to withdraw. But the United States didn't want him to withdraw. Washington basically wanted to drive him out, not have him withdraw. That led to the first Iraq war.[7]

Right after that war, the United States was in total control of the region. Saddam barely existed. Still, he launched a major assault against the Shiites in the south. The United States refused to block it. There was a big massacre of Shiites in the south, and the U.S. government didn't lift a finger, not even to block military helicopters.[8]

After that, Saddam turned against the Kurds in the north. But this time, the U.S. decided to protect them from Saddam Hussein. And suddenly, the reporting was quite different. The reporters went to the north. If you remember the television coverage at the time, they were appalled to find that atrocities were being carried out against the people whose children were blue-eyed and blond and just like us. We couldn't tolerate it. There was a big hue and cry. Finally, Bush established a no-fly zone. That's how it went for the Kurds in Iraq.

At the same time, in the 1990s, Turkish repression of the Kurds was extremely severe. Tens of thousands of people were killed, about thirty-five hundred towns and villages were destroyed, probably a couple of million refugees.[9] There was every imaginable form of torture. It was just a horrendous attack. And all of this was completely supported by the United States. Eighty percent of Turkey's arms came from the United States.[10] In fact, as the atrocities mounted, the arms flow increased. The atrocities actually peaked in 1997, and that same year Clinton sent more arms to Turkey than in the entire Cold War period combined.[11]

The press refused to report on any of this. It wasn't a secret. There are extensive reports from Human Rights Watch—they had a very good investigator there—and Amnesty International. You could find out what was happening, just not from reading the *New York Times*. The *Times* had a bureau in Ankara, of course, but it wasn't interested in covering this, especially the U.S. role. It wasn't the right story.

The Iraqi Kurds switched from unworthy to worthy. They might switch back. But all of this teaches a lesson. There is a Kurdish slogan: "Our only friends are the mountains." That's wise. The Kurds should not be deluded into thinking that just because the U.S. government is patting them on the head today, it won't be supporting another Halabja massacre tomorrow.

Incidentally, this worthy-unworthy distinction—I should have mentioned this—actually comes from George Orwell, who made a distinction between what he called people and unpeople.[12] People are those who count. Unpeople don't. You can do anything you like to them.

That was made vividly clear to me when I was speaking to a videoconference in London. The moderator brought up the horror in the West over the beheadings of journalists. "We are horrified," he said. "This is so hideous. We just have to do something about it." He was addressing a pretty liberal group. "We recognize that U.S., British, and Israeli atrocities are pretty awful, but even during the Israeli attack on Gaza, you didn't see things like beheadings."

Didn't you? Take the most recent Israeli attack on Gaza. In Shuja'iyya, people were picking up pieces of bodies to try to identify who'd been killed.[13] That was reported. But the moderator in London was correct: it didn't horrify the West. When we carry out atrocities like bombing people, leaving their body parts so scattered you can't even identify who they were, that's not a crime. It's perhaps a "mistake." Just like the "mistakes" that happen in the drone assassination campaign, which undoubtedly does worse things than beheading to its victims. Maybe it's a mistake, but it's not a crime. On the other hand, if ISIS beheads people, that offends us to the heavens. Those murders are horrendous, undoubtedly—but they're a tiny fraction of what we and our clients do.

The Iraqi Kurds have taken Kirkuk, a valuable center of oil, thus increasing the possibility of economic viability for an independent Kurdish state. Some Israeli and Turkish commentators have said it's inevitable. What do you think of that possibility?

It depends on what the master of the world decides. For the moment, at least, the United States is opposed, which means that the Kurds, though they have plenty of oil, can't sell it on the international market, because the United States won't allow it. Some oil undoubtedly gets sold, some gets leaked into Turkey. Israel is apparently purchasing some of it. But the Kurdish tankers are wandering around the Mediterranean, trying to keep from being too visible and trying to off-load the oil that they're carrying.[14] At this point, the Kurdish quasi-state can't even pay its officials.[15] They are not getting anything like enough revenue. Incidentally, all this is happening while the capital, Erbil, is full of high-rises, skyscrapers going up all over the place, tremendous wealth—the typical features of an oil state.

The Kurds are in trouble. They're landlocked. They have no access to the outside. Iraq refuses to provide them with the means to sell their oil through Iraq. They pretty much have to go through Turkey, and that's going to require U.S. support. So far that hasn't been forthcoming. So I don't think it's at all inevitable.

If you take a look at a map, the whole Kurdish region is sort of a unit. The biggest part is in southeast Turkey. Another part of it is in Syria. Assad has more or less been leaving that alone, so the Kurds have had a kind of semiautonomy during the Syrian disaster, but now they're under attack by Sunni jihadi forces, ISIS, Al Nusra, and others. The question is, Can the Syrian Kurds link up to Iraqi Kurdistan and maybe ultimately to Turkish Kurdish areas? There are very complicated negotiations going on between the Iraqi Kurdish leadership and the Turkish government. But the Kurdish areas in Syria are under the control of a group that is sympathetic to the Kurdistan Workers' Party (PKK), the Turkish guerrilla organization, which is a bitter enemy of Turkey and of the United States.

With the rise of ISIS and Salafi theology and ideology in the region, wouldn't this be an opportunity for rapprochement with Iran?

That's what the Iraqi government is calling for. Iran and the United States happen to be very much on the same side here. It's not the first time. Iran was strongly opposed to the Taliban and was very helpful to the U.S. government in its invasion of Afghanistan. In fact, in 2003, President Seyyed Mohammad Khatami made an offer to the Bush administration to put all of the contested issues on the table: Israel, nuclear weapons, everything. Let's discuss them all. The Bush administration rejected it.[16] We've decided Iran is an enemy. They're too independent. We won't tolerate that.

Incidentally, the same is true of Assad in Syria. The only major military force attacking ISIS happens to be Assad's quasi-government at this point, which is allied closely with Iran. Iran is also apparently sending arms, advisers, and probably troops to Iraq to support the Iraqi government against the ISIS assault. But the United States has insisted that the "international coalition" must exclude Iran and exclude Assad. So the main component of the coalition is Saudi Arabia, which is the main funder of ISIS and the ideological center for ISIS. It makes absolutely no sense.

The role of Turkey is central. Vijay Prashad, an author who teaches at Trinity College in Connecticut, recently said in an interview, "All evidence suggests that Turkey has allowed ISIS fighters, when they've been injured, to return into Turkey and to get treated in Turkey's hospitals."[17] The border is porous.

Yes. That's the border with Syria, and the ISIS fighters are just pouring across it. They're getting military support and medical aid. Turkey was under great pressure by Obama to join the great coalition. But they're plainly not joining. Turkey has an enormous military force of its own. If they entered the fight, they could wipe ISIS out in no time, just as Iran could. But Turkey is not interested—and Iran is not permitted.

Turkey is a NATO ally, a longtime recipient of U.S. military aid. It would seem that Washington would have the kind of leverage to exact what it wants in terms of sealing the border.

You would think so, especially given the U.S. backing for Turkey's vicious counterinsurgency operation against the Kurds. But the Turks don't simply follow orders.

Something extremely interesting happened in 2003, when the United States invaded Iraq. If you take a look at the map, it was obvious that the United States wanted to invade Iraq from Turkey. Those big military bases in eastern Turkey are right there on the Iraq border. They would have been a perfect base for the U.S. forces launching their attack. But the Turkish population was strongly opposed to the idea. Polls showed that more than 90 percent of Turks opposed the U.S. attack.[18] Not because they loved Iraq. They just didn't want to be part of U.S. aggression. To everyone's amazement, the Turkish military, which has tremendous power, permitted the Turkish government to follow the will of 90 percent of the population. That caused a furor in the United States. How dare Turkey refuse U.S. orders and pay attention instead to 90 percent of its population? The country was denounced in our press, which for the first time started reporting Turkish human rights violations. You hardly heard about these while they were going on in the 1990s, but all of a sudden we cared. Now we had to talk about how awful the Turks are.

The most striking case was Deputy Secretary of Defense Paul Wolfowitz. In the media, he was called the "idealist in chief" of the Bush administration.[19] He was the deeply moral person, over-the-top idealistic. He bitterly condemned the Turkish military because they didn't force the Turkish government to accept U.S. demands. He even insisted that the military apologize to the United States and make it clear that it would never commit another crime like this.[20]

This was going on just as the government and the media and the intellectual community were orating about the U.S. dedication to "democracy promotion." If you want to be a prestigious intellectual or journalist, you have to maintain completely contradictory ideas at the same time and not notice it.

Orwell's "doublethink."

Yes, that's Orwell's definition of doublethink: the ability to have contradictory ideas in your mind and accept them both without noticing. That's practically a requirement in the intellectual world.

Since the founding of the Turkish republic in 1923, the military has been the dominant institution in that country. How has Erdoğan been able to sideline the military?

He instituted a big purge of the top military, and he got away with it. The military has been reduced in its power over the government. How much is not clear, but substantially. That was one of Erdoğan's major achievements in the first half-decade of the millennium.

Minorities in the Middle East—the Yazidis in Iraq, Armenians in northern Syria, and other groups—are getting hammered. What can be done to protect them?

There is a framework of international law that, in principle, everyone accepts. It's spelled out in the U.N. Charter, an international treaty that the United States has ratified, which, according to the U.S. Constitution, makes it the supreme law of the land.

The Charter, specifically Article 39, says that the Security Council has to determine if there is a threat to peace—for example, the massacre of the Yazidis. Furthermore, the Security Council—and the Security Council alone—can authorize the use of force in a case they determine to be a threat to peace. Aside from that, there's an absolute ban against the threat or use of force except as direct self-defense against armed attack, which is irrelevant here. So that's the basis for protection.

But the United States, Britain, Israel, and other clients are rogue states, states that disregard international law. The U.N. Charter doesn't apply to them. They have a monopoly on force, or they want to have a monopoly on force, and they use it as they like. That restricts the options for how this problem can be dealt with.

In a law-abiding world, our government would ask the Security Council for a resolution declaring that there is a serious human rights situation and threat to peace in areas controlled by ISIS, and then ask the U.N. to authorize the use of force to deal with that threat. That use of force should primarily involve regional actors, including, of course, Iran.

But that's not what happens. Indeed, there is no mention in the press that there could be a lawful way to deal with this issue. That's beyond the consciousness of Western intellectual culture. The concept that we could act as a law-abiding state is unimaginable. If you mention it, people don't know what you're talking about. It's not an option. So it doesn't arise. What's done is what the master decides should be done.

In the September 18, 2014, referendum in Scotland, the vote was 55 percent to 45 percent to stay with the United Kingdom. What are its implications for Kashmir, the Armenians in Nagorno-Karabakh, and the Kurds in Iraq?

There are conflicting tendencies at work in Europe. For the last couple of hundred years, Europe was the most savage place in the world. Europeans had no higher goal than to slaughter one another. During the Thirty Years' War, in the seventeenth century, about a third of the population of Germany was wiped out. And then you had the two monstrous wars in the twentieth century. By 1945, the Europeans had comprehended that the next time, it would all be over, because the level of destructive technology had reached a point where they couldn't play that game anymore. And they did change their behavior. France and Germany, which had been slaughtering each other for centuries, moved toward peaceful reconciliation. Then the European Union started to integrate. Free movement in the European countries is a generally positive development, reducing the emphasis on national borders and leading to greater interactions among people who ought to be cooperating, not fighting each other.

But there are other, countervailing tendencies. Democratic participation has severely declined. Decisions over the European economy are made by bureaucrats in Brussels, mainly under the influence of the German Bundesbank. The opinions of people in Europe are mostly disregarded. There have been times when this has become almost surreal. In 2011, the prime minister of Greece, George Papandreou, made the mild suggestion that the people of Greece should be allowed to have a referendum to decide whether they would accept the harsh austerity measures decreed by the bankers in Brussels.[21] The West was just outraged. The press, intellectuals, and others denounced Papandreou for daring to ask the population whether they should follow the orders of the bureaucrats and the bankers.

It's led to a complicated reaction in Europe. Some of it is frightening. There is a right-wing reaction—in some places neo-Nazi, in other places just horribly right-wing—that is a response to the loss of democratic participation. But there is another reaction that, at least in my view, is healthier, and that's an impulse toward regionalization in opposition to the centralization of the European Union. So in a number of parts of Europe, people are calling for autonomy. Scotland is one case. Catalonia is another. It's happening in the Basque country, in parts of France, and elsewhere.

Europe is a complex of cultures, languages, history, a complicated tapestry. But one of the things that's happening is the rapid destruction of local languages. They're dying very quickly because the nation-state system imposed national languages instead. In Italy, for example, there are plenty of people who can't talk to their grandmothers, because they speak a different language. But there's a countertendency toward reviving regional languages and regional cultures. I think the Scottish referendum is part of this.

The same issues arise globally. In the Middle East, the state system was simply imposed by imperial power. The lines of the states have nothing to do with the people of the region. Take Iraq, for example. The British established modern Iraq in their interests, not in the interests of Iraqis. So they took the region around Mosul and added it to Iraq because Britain wanted to have the oil and keep it from Turkey. They set up the principality of Kuwait to keep Iraq from having free access to the sea, so it could be better controlled. The Sykes-Picot treaty, between France and Britain, assigned Syria and Lebanon to France, and Iraq and what was then Palestine to Britain. That was for their imperial interests. It had nothing to do with the people. The lines make no sense from the point of view of the people.

The Ottoman system, which had preceded this, was ugly

and brutal, but at least it recognized local autonomy. So during the Ottoman period, you could go from Cairo to Baghdad to Istanbul without crossing a border. It was porous, sort of like the European Union today. And that fits the nature of the region much more accurately. Partly out of corruption and incompetence, the Ottoman rulers allowed considerable autonomy, even to some parts of cities. The Armenians could run the Armenian community, the Greeks could run the Greek community, and so on. They lived in a kind of harmony. That was broken up by the imposition of the state systems.

This is true all over the world. Take a look at Africa. Almost all of the conflicts there trace back to the establishment of borders by the imperial powers—England, France, Belgium, to a lesser extent Germany—which took no account of the nature of the populations, just drew the boundaries where they wanted them. Naturally, that leads to conflict. There is every reason to hope, I think, that those borders will fade away.

5

LIVING MEMORY

CAMBRIDGE, MASSACHUSETTS (JANUARY 23, 2015)

What's the significance in the dramatic fall in the price of oil? There's article after article saying this is great for consumers, gas is under two dollars a gallon, people will be driving more, they will have extra money in their pockets.

It's an incredible moment, when you look at it. The business pages and the press are lauding the prospect that we can devastate the world for our grandchildren. The headline ought to be "Let's Destroy the Possibility of Our Grandchildren Having a Decent Life."

It's pretty dangerous, and it's getting worse every day. The latest concern is that there might be an explosion of methane from the melting of the Arctic and the permafrost. If that happens, some of the predictions are very dire.[1]

The price of oil is already too low. Oil should be priced much higher on the U.S. market, the way it is in Europe, to try to discourage excessive use of fossil fuels, which are destroying the environment.

The evidence for climate change seems to be incontrovertible and should be totally noncontroversial. Newspapers announce, "Ocean Life Faces Mass Extinction," "2014 Was World's Hottest Year Since Record Keeping Began in 1880," "10 Hottest Years Have Occurred Since 1997."[2] Yet the response from the political class and the owners of the economy seems lukewarm, tepid, and merely cosmetic at best.

There was just an interesting poll done by PricewaterhouseCoopers. It was released at the meeting of all the big shots in Davos, Switzerland. They polled the CEOs of corporations about what they considered to be the significant issues they face. What they cared about most was profits—what's the growth situation like, will we have enough low-paid workers? Climate change was at the very bottom, a minor concern way out on the margins.[3]

It's not that they're bad people, but there is an institutional pathology. If you're the CEO of a major corporation—which, incidentally, means that you have enormous influence on the political system—you simply don't care about what happens to the world in the future, including to your own grandchildren. What you care about is profits tomorrow. It's an institutional imperative.

There is a Yanomami shaman leader named Davi Kopenawa. There are about thirty to forty thousand Yanomami in northern Brazil and southern Venezuela. He says, "The white people want to kill everything. They will soil the rivers and lakes and take what is left. . . . They do not think that they are spoiling the earth and the sky, and that they will never be able to re-create new ones. . . . Their thoughts are constantly attached to their merchandise. They relentlessly and always desire new goods."[4] Many indigenous people—I'm not saying all across the board, of course—clearly have a different relation to nature.

That's pretty much true around the world. In southwestern Canada, the indigenous people, First Nations, are leading the struggles, mobilizations, and legal efforts to prevent the extremely dangerous expansion of highly destructive fossil fuel use.

In the Amazon, indigenous people are in the forefront of efforts to prevent overuse of fossil fuels and other resources and restore some kind of balance with nature. In fact, two of the Latin American countries with the largest indigenous populations—Bolivia and Ecuador—have been in the lead in trying to establish what they call "rights of nature." It's even a constitutional provision in Bolivia.[5]

The same is true in Australia. And in India, where tribal people are trying to protect resources. These communities have lived in some kind of balance with nature for a very long time. I don't want to make it sound as if they live in some kind of utopia, but at least they've long had some concern for a balance with nature. And the capitalist, imperialist invaders did not have that concern. You can see it in the poll of CEOs, which is perfectly typical of the attitude of imperial powers that just want to ravage the world and take it for their immediate use.

You had some contact with indigenous groups in the Colombian rain forest.

I have spent some time in southern Colombia, which is a highly embattled region. Campesinos, indigenous people, and Afro-Colombians are under constant attack by paramilitaries, by the military, and also by the guerrillas, who used to be connected to the local populations but have now become another army preying on the peasants. Also, there is what we call "fumigation," which is chemical warfare that destroys virtually everything. Theoretically, it's aimed at coca production, but in fact it destroys all kinds of crops and livestock. You walk through the villages and see children with gruesome sores on their arms. People are dying.

One time, I went with some Colombian human rights activists to a remote village. A group of campesinos and indigenous peoples there are trying to preserve their water supplies. There is a mountain with a virgin forest that is their source of water, and also has meaning in their cultural life. It's being threatened by mining. The community has quite sophisticated, thoughtful plans about how to preserve the area's hydrological and other resources, but they're fighting against powerful forces: the mining companies, the government, the multinationals. It's a battle. And it's very violent. The first time we tried to visit the area, we weren't allowed because there was too much killing going on. We were able to get through the second time.

You have a family connection, as well. Can you talk about that?

Yes. I was there because they were dedicating a forest on the mountain to my late wife, Carol. The villagers all participated in the dedication. It was a moving ceremony. There were shamans and so on. It was pretty dramatic.

In Power Systems, *our previous book of interviews, you said that Latin America "has shown increasing independence in international affairs."[6] Is that trend continuing?*

It is, definitely. I think it's probably the major factor behind Obama's move to what we call "normalized relations" with Cuba, meaning a partial end to the attack on Cuba that's been going on for fifty years. I suspect part of the reason is that the United States has been seeing increasing pressure from the rest of the hemisphere on this issue. Back in the early 1960s, when the United States kind of ran the show, Americans demanded that Cuba be excluded from the hemispheric organizations. Now, as Latin America has become more independent, more free of U.S. dominance, it has increasingly insisted that Cuba be allowed back into these groups.

When Obama announced the shift in policy vis-à-vis Cuba, I didn't see any mention of the extensive terrorist campaign, trade embargo, and economic warfare the U.S. government carried out against Cuba. And no mention, of course, of reparations or compensation.

There was one mention of the terrorist war, and it was about the silly CIA pranks, trying to burn Castro's beard or something like that.

Poison pens.

We're allowed to make fun of that. But not to mention the fact that John F. Kennedy launched a major terrorist war against Cuba. His brother Robert Kennedy was placed in charge of it.

It was his highest priority. And the goal was to bring "the terrors of the earth" to Cuba. That's the phrase that was used by Arthur Schlesinger, JFK's Latin America adviser, in his biography of Robert Kennedy.[7] And they did bring the terrors of the earth: blowing up petrochemical plants, sinking ships in the harbor, poisoning crops and livestock, shelling hotels (with Russian visitors in them, incidentally). It went on for years. It was one of the factors that led to the missile crisis, which almost caused a nuclear war. When the missile crisis ended, Kennedy instantly relaunched the terrorist war, which went on in various forms into the 1990s. None of that gets discussed. Actually, the first oral history, featuring testimony from victims, appeared in 2010. It was written by a Canadian researcher, Keith Bolender, but it hasn't been read here.[8]

Obama's message, echoed in the media, is that our efforts to bring democracy and freedom to Cuba have not succeeded. Although they were all benevolent in intent, they haven't worked. It's therefore time to try a new method to achieve our noble goals. That's Obama's description of fifty years of massive terrorism and economic strangulation, strangulation so extreme that if, say, a European manufacturer of medical equipment used a little piece of nickel imported from Cuba, his business would be banned from international commerce.

So that was our benevolent effort to bring democracy and freedom to Cuba. Not to the dictatorships that we support. We don't make benevolent efforts there, somehow.

The U.S. war on Cambodia was called a sideshow, the main event being Vietnam. The sideshow to the sideshow took place in landlocked, mostly rural Laos. In March 1970, on your way to Hanoi, you were delayed for a week in Vientiane, Laos. You wrote about that in the New York Review of Books. *The essay was later published in* At War with Asia.[9] *I was struck by your descriptive journalistic writing—clear, terse sentences.*

You had a very moving experience with Fred Branfman, who passed away in September 2014. He had been in Laos for many years, and spoke Laotian. You went with him to a refugee camp outside of Vientiane.

I didn't know him at the time, but we met soon after I arrived. He had been trying for some time to get some Western exposure for the atrocities in Laos. He was one of the very few people working in Laos—along with Walt Haney and a couple of others—who had discovered the crimes that were being committed, which were really shocking. The book he produced, *Voices from the Plain of Jars*, is the result of his work with victims of the horrific air war taking place.[10]

The bombing of Laos started in the mid-1960s, and intensified in 1968. The Plain of Jars was a remote area of peasant villages. Most of the villagers probably didn't even know they were in Laos. They were subjected to years of extremely intensive bombing. People were living in caves, trying to survive. One should really read the testimonies in Fred's book to get a picture of it.

I was in Laos for a week, thanks to the boredom of an Indian bureaucrat. Bureaucrats have nothing to do except make life difficult for people. This guy was in charge of U.N. flights from Vientiane to Hanoi. There was one flight a week, through a special protected corridor. When you flew, you saw jet planes all over the place on their way to bomb whoever. For some reason, the bureaucrat decided not to let us go the first week. It kind of amused him. So I stayed in Laos, which turned out to be a very good thing, because I learned a lot. I spent most of the week with Fred, not just in the refugee camp. I went to the village where he had lived, and met some of his many contacts.

You don't name Fred in your article. You said you were "in the company of a Lao-speaking American."

He did not want to be identified at that time.

Fred wrote an article about his friendship with you.[11] I don't want to embarrass you, but he said that you broke down when you met those villagers and heard the stories of living through the U.S. bombing.

Laos was the first time—there have been many since—when I saw firsthand the effect of massive atrocities on the victims. I had been in the U.S. South during the civil rights movement, which was bad enough, but I hadn't had exposure overseas before Laos. And, yes, it was a shattering experience.

In the foreword to the second edition of Voices from the Plain of Jars, *historian Al McCoy writes that approximately twenty thousand civilians have been killed or maimed by unexploded cluster bombs since the bombing ended—and those numbers continue to rise.[12]*

That's correct. I've written about it, too. These are tiny little bomblets, as people call them. Children pick them up, thinking they're toys, and are blown up. They also explode and maim farmers who hit them with a hoe.

They're all over the place. A British demining team has been working to remove the bomblets, but the area is saturated with them. It's a massive undertaking. And very limited resources have been devoted to the effort by the United States, which is responsible for the situation, of course.

McCoy suggests that Laos was a test case for future U.S. wars, particularly the extensive use of airpower.

Fred also talked about that. We have other test cases, as well, which are pretty remarkable. Researchers at the Seton Hall law school, who published a detailed study of the Guantánamo torture system, point out something quite interesting. There was a part of the Cheney-Rumsfeld torture system in Guantánamo that the military called "the battle lab."[13] It was essentially a laboratory of torture. The lab was supervised by medics, and its purpose was to determine the most effective techniques of torture. Let's figure out how much torture—psychological, physical, medical—can be applied.

In fact, if you take a look at the Senate report on the torture system, it asks one question: Did torture work? And it claims torture didn't work, so therefore it was bad.[14] The commentary has been pretty much the same: torture didn't work, so we shouldn't do it.

When they say torture didn't work, it means it didn't stop terrorist acts. But was that the purpose of it? Probably not. The initial purpose of the Cheney-Rumsfeld torture system seems to have been to try to extract some kind of claim—true or false, it doesn't matter—that would justify the war in Iraq. They were trying to find some kind of evidence that there were connections between Saddam Hussein and Al Qaeda. When they didn't find it, they called for more torture. Finally, because people under torture will say anything, they claimed they got some evidence. Apparently, that was the primary goal. And it was achieved.

We are now in a new era of terrorism, sparked by the 2015 attacks in Paris. There was a lot of commentary about them as an attack on freedom of speech, French values, the West in general.

You're speaking of the attack in which *Charlie Hebdo* journalists were killed. One of the most interesting comments about that was by the leading civil rights lawyer Floyd Abrams, well known for his vigorous defense of freedom of speech. He castigated the editors of the *New York Times* because they didn't publish the *Charlie Hebdo* cartoons ridiculing Muhammad that had elicited the attack. He said, If you really want to serve the highest values of freedom of speech, you should publish those cartoons. He said, This is the right way to honor "the most threatening assault on journalism in living memory."[15] And he's right. But the category of "living memory" is very carefully crafted to include anything they do to us and to exclude anything we do to them.

If you go beyond "living memory" to the actual world, there are many such attacks on freedom of speech, some of them quite similar. For example, NATO—that is, the United States—bombed a radio and television station in Serbia to knock it off the air and killed sixteen journalists, actually more than died at *Charlie Hebdo*.[16] Why? Because it was broadcasting information supporting the government that we were attacking.

The coverage of the Fallujah attack in November 2004 is also quite interesting. The general hospital there was occupied by American Special Forces, a war crime. The command was asked about it by reporters. They said the hospital was a legitimate target because it was producing propaganda—namely, casualty figures—so therefore we had to occupy it.[17] Is that an attack on freedom of expression?

France is a particularly striking case because the national government actually has laws that grant the state the right to determine historical truth and to punish deviation from state edicts. That in itself is a pretty extreme attack on freedom of expression. So, for example, they've closed Basque nationalist newspapers in southern France, which they thought were disruptive of public order because they were calling for Basque independence.[18]

The record is replete with double standards and hypocrisy. Look at Saudi Arabia, our great democratic ally. They're flogging a blogger, Raif Badawi.[19]

I don't even think the phrase "double standards" is appropriate. There is a single standard: If they do it to us, it's a horrible crime. If we do much worse to them, it's a noble endeavor. That's a single standard. And it's maintained with remarkable consistency and dedication.

Actually, there is an even more general principle. The more that we can attribute a crime to some enemy, the greater the outrage. The more we're responsible—and, incidentally, can therefore do something about it—the less the concern, going all the way to indifference or, even worse, denial. That's a principle that's applied with overwhelming consistency.

So then, where do you come down on the issue of freedom of speech? For example, the publication of cartoons—not just about the prophet Muhammad but also about the pope or the Catholic Church—that offend the sensibilities of some people. Are you an absolutist on free speech?

I'm not an absolutist. I don't think we can take absolute positions on any moral or ethical principle. But I think it's a very high value to be defended. I think the U.S. Supreme Court did set a standard that's pretty reasonable back in 1969, in *Brandenburg v. Ohio.* Speech should be protected unless it's intentionally directed at producing an imminent criminal act.

Publishing cartoons that ridicule and humiliate people is at the level of a stupid adolescent prank. And it's really vulgar when you're attacking people whom you're grinding under your jackboot. It's one thing to make fun of the powerful. If you want to do that, okay. But to make fun of people you're crushing, to ridicule them, is particularly obscene.

If you take a look at the people the *Charlie Hebdo* cartoonists were ridiculing when they published cartoons of Muhammad, it's poor, oppressed people, North Africans mostly, who fled regions that France has devastated. France has had a hideous record of extermination and violence in those regions for well over a century. A lot of people from there have ended up in the suburbs of Paris, the *banlieus*, where they live in slums and are subjected to repression, contempt, degradation, humiliation. And then you publish cartoons ridiculing them. Very funny.

But even though it's an infantile vulgarity, I think it should be protected.

Jews were singled out for killing in the attack on the kosher supermarket in France. Benjamin Netanyahu and other members of his government have used that to urge French Jews to move to Israel.

Yes. The attack on the kosher supermarket was a couple days after the *Charlie Hebdo* incident. They took hostages there and then killed four people.[20]

There is undoubtedly anti-Semitism in France. It's real and it's dangerous. It's not anywhere near the level of Islamophobia, though—and the idea that Jews aren't safe in France is an illusion. Take terrorism in general. How many people in the United States are killed by terrorism? More people here are killed by accidents at home than by terrorism, a lot more.[21]

Terrorism exists, and it is serious. And there are several ways to deal with it. The main way is to stop participating in it. Obama's global assassination program is the most extreme terrorist campaign in operation by far. It's a campaign aimed at killing people who are suspected of perhaps planning to harm us someday, and anyone else who happens to be around them. That's pure terrorism, on a massive scale.

Incidentally, Israel does exactly the same thing, killing people whom they suspect of plotting acts against Israel. Israel bombed Syria on this pretext, for example. The target was apparently a man whose father had already been killed. The attack also killed some Iranians, it turned out.[22]

Suppose Iran were to murder people in Israel whom it suspects of plotting against Iran. That's not far-fetched, because Israel does plot against Iran, even calling for attacks. So suppose Iran decided to kill Israelis in Israel. Would that be fine? No. We would go to war. But when we do it, on a massive scale, killing thousands of people, it's fine.

If we were to cut back on these actions, it would sharply reduce the amount of terrorism in the world. But it also would have another effect. It's well known by now that our drone attacks generate terrorism in response. Incidentally, that's one of the things that Fred Branfman studied. He provided a lot of evidence that, even at the highest level of the U.S. government, people understand that these attacks are creating jihadis. So if we cut down our own terror operations, we would also cut down the kind of terror that enters "living memory"—namely, attacks against us.

Take a look at the studies that have come out. People have been incited to jihad by seeing the torture at Abu Ghraib, the humiliation and degradation and repression the prisoners suffered daily. That doesn't justify terrorism, of course, but it helps explain it. And it tells you what to do if you want to cut it back: eliminate, reduce the actions—our actions—that are helping to generate it.

Two of the attackers in Paris were of Algerian descent. Robert Fisk reminded us in a recent article that over one million Algerians were killed by France during the war of independence, between 1954 and 1962.[23] That's almost one in ten.

He's quite right. That's not living memory for us, but it's living memory for the victims. And they can remember even further back, way back into the early nineteenth century. When the French invaded Algeria, the explicit goal was to exterminate the population. And they did a pretty good job of it, not just in Algeria but in all their colonial areas.

Let's move on to a less controversial topic. You came to a conclusion about God as a result of observing something your paternal grandfather did.

My father's family was extremely Orthodox, ultrareligious, especially my grandfather, who had come from Eastern Europe and maintained the sort of semi-medieval characteristics typical of the Eastern European rural Jewish community. I remember we would visit for the Jewish holidays. On one of the holidays, Passover, I noticed that my grandfather was smoking. I knew that making a fire was not allowed on the Sabbath, and that the Talmud says there is no difference between the holidays and the Sabbath except with regard to food. So you're allowed to make a fire to cook on the holidays, but nothing else. And my father told me that my grandfather had decided that smoking was a form of eating.

So I realized that he thought God was so stupid that he couldn't see through this. And when I thought about it, it seemed like practically all of organized religion was based on the assumption that God was so stupid that he wouldn't notice you violating his commandments. Almost nobody can live up to the commandments, so you find all kinds of trickery to get around them. And if that's your conception of God, from a ten-year-old's point of view, it didn't seem worth pursuing.

When did you become convinced that there was no God?

I never became convinced, because I don't even know what the question is. What is it that there isn't? There is no coherent answer I know.

At a talk you gave at Princeton, you recalled that one of the things that got you interested in linguistics was that you realized the Bible was mistranslated.

I was told about it. I was studying Arabic in college with Giorgio Levi Della Vida, an Italian anti-fascist émigré—a leading scholar, though I didn't know it at the time. We became good friends later.

This was at the University of Pennsylvania?

Yes. He mentioned to me that the first sentence of the Bible was misvocalized. What it says is *"Bereshit bara,"* which is ungrammatical, and it's translated, "In the beginning God created." But the actual Hebrew should be vocalized as *"Bereshit bro"* and translated as "At the outset of the creation, there was chaos." It had gone for a thousand years with nobody noticing.

And you inferred something from that?

That there is a lot to learn.

You've said the Talmud is your ideal text. Why?

If you look at a page of the Talmud, in the middle of the page is a passage taken from the Mishnah, a book of laws. And then around it there are running commentaries. In the upper right-hand corner there is a commentary from someone, and then in the upper left-hand corner a commentary from someone else. Ninety percent of the page is commentary about this line in the middle. If only one could write footnotes like that, it would be fantastic.

6

FEARMONGERING

SANTA FE, NEW MEXICO (MARCH 18, 2015)

Senator Tom Cotton of Arkansas recently announced that we have a great deal to fear from Iran, because "they already control Tehran."[1]

And if you read the *Washington Post*, you will discover that Cotton—who has a real pedigree, he even graduated from Harvard—is positioning himself to be the future foreign policy specialist of the Republican Party, taking on the mantle of John McCain and Lindsey Graham.[2] He also has other interesting ideas. I don't know how much you've followed his career, but when he was running for the Senate in Arkansas, he warned that the Mexican drug cartels are linked to ISIS and are working to send terrorists across the border, where they can kill citizens of Arkansas. And, of course, all of this was the fault of President Obama for leaving a porous border and so on.

We've seen this kind of fearmongering elsewhere, too. In

Israel, Benjamin Netanyahu warned the electorate that Arab citizens of Israel are being driven to the polls by leftists with support from foreign governments, all in an effort to undermine his policy of defending Israel from terrorists.[3] That combination of fearmongering and racism works quite well, unfortunately.

Electronic Intifada cofounder Ali Abunimah says Netanyahu is good for the Palestinians. Why? Because he's very clear: no Palestinian state, no compromise. Haaretz *correspondent Amira Hass sees only cosmetic differences between the two major Israeli parties, and she says that the now moribund two-state solution is actually a ten-state solution, "a bunch of bantustans," she says, "inside the West Bank."* [4]

I would put it a bit differently. Israel is carrying out a perfectly reasoned, intelligent program intended to integrate into Israel everything in the West Bank that might be of any value, but to exclude the Palestinian population. The areas that Israel is taking over don't have many Palestinians, and those who remain are largely being expelled. So, no one state, no "demographic problem," and Palestinians lose everything. That's the actual alternative to a two-state approach. No one has made a meaningful case that there could be one state. The likely alternative to a two-state settlement is the policies I've just described, which are now being implemented.

As long as the U.S. government continues to support this, there is no reason to expect Israel to stop, whether Netanyahu is in charge or anyone else. The racist and extremely alarmist rhetoric of Netanyahu is not shared by the other Israeli parties, so there are some differences in policies. But Amira Hass is correct in saying that they are not fundamental ones.

Years ago, the philosopher John Dewey said, "As long as politics is the shadow cast on society by big business, the attenuation of the shadow will not change the substance."[5] How would you evaluate that shadow today?

Dewey's comment was accurate. Just look at the moves toward privatization of Medicare in Paul Ryan's budget. Medicare is the one part of the health care system that more or less works—because it's *not* privatized. Its inefficiencies and costs, such as they are, are due to the fact that it has to work through the highly inefficient, bureaucratized, privatized system we have in the United States. So moving to privatize Medicare is saying, "Let's undermine the one system that more or less works."

Cutting back Medicaid; cutting back food stamps; making tens of millions more people uninsured by repealing the Affordable Care Act—not wonderful legislation but, nevertheless, an improvement over what was there before—and, at the same time, giving more money to the wealthy: that's the Republican radical insurgency for you. Their one consistent policy is to do anything they can to enrich the wealthy and powerful, on the one hand, and attack the general population, on the other.

Since you can't win votes on that platform, the Republican Party managers have had to obfuscate their position. They've turned to sectors of the population that have always been around but had never been mobilized into a significant political force. One of these is Christian evangelicals, a major part of the Republican Party base today. Another is nativists, people who are afraid that all these others are taking our country away from us. The white population will become a minority pretty soon, and for extreme ultranationalists, that situation can't be tolerated. There are sectors of the population who are so frightened that they have to carry guns into Starbucks. Who knows what might threaten them there? In fact, there's been legislation debated in Ohio that would allow guns to be brought into daycare centers.[6] Maybe some of the three-year-olds were trained by ISIS. Who knows?

These are not small parts of the population. A lot of people can be mobilized on those issues and not notice that the policies their leaders are pursuing are attacking them. It's a very strange country in many ways.

If you look at attitudes regarding Obamacare, they've been pretty negative. Most people have been opposed to it, even though for years much of the population has been strongly in favor of national health care. Of course, Obamacare isn't national health care, and some of the opposition to it—we don't know how much, because that question never gets asked in the polls—is because it didn't go far enough. But a lot of the opposition is the kind of thing you see reflected in this famous town hall comment, where somebody said, "Keep your government hands off my Medicare."[7] People don't understand what the government actually does. This is a remarkable triumph of propaganda, if you think about it, especially considering how vital health care is to everyone's life.

The desire to enrich the wealthy and powerful at the cost of everyone else isn't restricted to Paul Ryan, of course. You were just in Argen-

tina and met with some activists from the Podemos movement in
Spain. What were your impressions?

This was an international conference of activists from around
the world, mostly from South America, but some from Podemos
in Spain, some from Syriza in Greece, and others.

For five hundred years, since the early European conquests,
South America has been dominated by foreign powers. The
typical government structure was a small, extremely wealthy,
Europeanized, mostly white elite ruling over vast misery and
poverty. The elites were oriented to the outside world. They had
their second homes on the Riviera, sent their money to Zurich,
and so on.

South American governments used to be the most accepting
of the neoliberal structural adjustment policies of the World
Bank, the International Monetary Fund, and the U.S. Treasury
Department. Naturally, these countries were the ones who suffered the most. But in the last ten or fifteen years, they've pulled
out of that pattern, moving out of our control. This is a major
change in world affairs. That's why the conference was in South
America—and why the organizers invited participants from
Spain and Greece, which have been hit especially hard by those
policies.

The savage economic program that Europe has been subjected to has seriously undermined democracy. It's been particularly devastating for the weaker, peripheral countries.
The policies of austerity under recession are economically
destructive—even the International Monetary Fund says they
make no sense from an economic point of view.[8] But they make
some sense from the point of view of class war. They're enriching

the big banks while dismantling social programs. I think that's the purpose of the policies: dismantling the social-democratic welfare state, Europe's major achievement after the Second World War.

Naturally, there has been a strong reaction against them—first in Greece, which has suffered the most. Greece called for a restructuring of its debt, a delay of debt payments, which might extricate the country from this artificially created disaster. But the German banks, which are basically responsible for the crisis, reacted in an absolutely savage way to prevent Greece from taking such steps.[9] This refusal is particularly ironic because in 1953, Germany was permitted by other European countries to cancel its major debts.[10] Indeed, that was the basis for the German recovery, and the reason Germany became the dynamic center of Europe. What's more, Germany practically destroyed Greece during the Second World War. So now, Greece is asking for a limited element of what Germany was granted in 1953, and the powers in Germany, the Bundes-bank, are just flatly refusing.

Greece is a pretty weak country, but Spain is bigger, with a more powerful economy. In the past few years, a new political party has developed in Spain, Podemos, which is dedicated to reversing the austerity programs and rebuilding the social economy. In Spain, as in Greece, the criminals who caused the crisis were the banks, the Spanish banks and the German banks. And in Spain, too, they want the population to pay the costs.

Notice that none of these people believe in capitalism. In a capitalist society, if, say, I lend some money to you—and since I know you, I know it's a risky loan—I charge you a high enough interest rate to account for the risk. If at a certain point you can't pay, it's my problem. That's how it works in a capitalist society, but not in the societies we live in. Here, my problem becomes my neighbors' problem, too. The neighbors didn't take on the risk, but they're made to pay for it by bailing me out. That's the way our system works. It's radically anti-capitalist. It makes sense on class warfare grounds, but has no resemblance to capitalism or free markets.

You grew up in the 1930s, at a time when solidarity meant something. There was mutual support, there was an active labor movement. What's it going to take to rekindle that spirit of solidarity?

Go back to the 1920s. The labor movement had been destroyed. There was practically nothing left of it. David Montgomery, a leading labor historian, wrote a book about this period called *The Fall of the House of Labor*.[11] There had been a lively, vibrant, active, fairly radical labor movement in the United States, but it had been crushed. The business classes had the support of state power. They were able to crush and destroy the labor movement. But it revived.

In fact, in the 1930s, the labor movement moved to the forefront. There were sit-down strikes and organizing by the Congress of Industrial Organizations. The Roosevelt administration was sympathetic to an extent, and was willing to accommodate some of the demands from the public, which were spearheaded by the labor movement. The result was New Deal legislation, which was very beneficial to the population and to the economy.

And the labor movement can revive again, as can other popular movements. There is a reason for being hopeful here:

the quite positive changes that have taken place since the 1960s. In many ways, today's society is much more civilized than what we had back then. When I started giving public talks during the early days of the Vietnam War, they took place in somebody's living room or in a church and were attended by only a handful of people. None of us who were involved ever could have imagined at the time that there would be a major anti-war movement a couple of years later. But there was.

And the same has happened on other fronts, as well. Lots of issues which you could barely discuss back in the 1960s are now accepted and taken for granted: women's rights, gay rights. There was no concern for environmental issues in the 1960s. Now there's substantial concern. And I think that's the starting point for re-creating the kind of solidarity, mutual aid, cooperation, dedication, and commitment that is necessary today.

We can't overlook the fact that we're at a moment of human history that is entirely unique. For the first time in human history, the decisions we make will determine whether the species survives. That has not been true in the past. It's very definitely true now.

7

ALLIANCES AND CONTROL

*I want to start with a George Orwell essay from 1946, "Why I Write."
He says, "My starting point is always a feeling of partisanship, a sense
of injustice. . . . [I write] because there is some lie that I want to
expose."[1] What's your starting point?*

That's very hard to say. A lot of my work is scientific work,
and the starting points are problems, puzzles, a desire to
understand—to borrow the title of my last book—what kind
of creatures are we.[2] The rest is about things happening in
the world, domestically or internationally, that seem to be
misleadingly or falsely described and are significant enough
to merit close attention.

*You, Stephen Hawking, and others signed a petition warning of an
artificial intelligence (AI) arms race.[3]*

I think the petition was initiated by Max Tegmark, a fine physicist here at MIT. The concern primarily is with automated military systems, which are extremely threatening. Automated systems can do many technically impressive things, but there are times when judgment matters, and they don't have it. If missile and nuclear systems are automated, we can expect errors, and those errors might well be lethal if there's no human intervention. As the systems become more and more automated, they are less and less controllable.

The petition also speaks of beneficial artificial intelligence. What might be some of the benefits of AI?

It would be nice to have a robot clean your house, cook your meals, drive your car. Robots can do useful—in fact, sometimes very useful—things: for example, replacing humans in extremely dangerous work involving radioactivity or other risks. Or, for that matter, just routine and boring work. To my mind, systems that improve our capacity to live a full, decent, and productive life are all to be welcomed.

We're witnessing the greatest human migration in Europe since the end of World War II. What are your thoughts about this unfolding human catastrophe?

Unfortunately, there are a number of catastrophes, though we shouldn't exaggerate the scale of them. Kenneth Roth at Human Rights Watch recently pointed out that if you consider the total number of refugees who are likely to try to make it to Europe, it is well under 1 percent of the population.[4]

Of the global population?

No, of the European population. For some countries, such as Germany, the influx is very welcome economically and socially. The refugees, especially the ones from Syria, are educated middle-class people with skills. Germany has a demographic problem. The society is not reproducing, so there's a shortage of young, skilled people. That is one of the reasons why Germany is taking a fairly welcoming position compared to the other countries in Europe.

Some other countries have also welcomed refugees, right?

Yes. Lebanon, for example, a small, poor country: by now, maybe a fourth of its population are refugees. Iran takes refugees. Jordan takes huge numbers. Turkey has taken an enormous number of Syrian refugees. Syria itself was accepting many refugees, until it started imploding.

There are also countries that generate refugees. The U.S. invasion of Iraq set all sorts of crises in motion, including the rise of ISIS, but it also created numerous refugees. Nobody knows exactly how many, but maybe one or two million, along with a couple of million displaced people inside the country.[5] Refugees are fleeing Iraq. They're fleeing Afghanistan. They're fleeing Libya, after we smashed up that country.

So, there are countries that accept refugees, there are countries that generate them, and then there are countries that generate them but refuse to accept them—like us. We may take in a few thousand refugees from the region, but nothing like the number created by the actions that we undertook.

You can say the same for Britain and France, on a smaller scale.

Remember, refugees are not coming because they want to. In fact, the United Nations has appealed for humanitarian aid to help refugees stay where they want to be, near the countries of their origin. But they're only getting about half of the aid that they requested.[6] The most constructive and humane way to deal with refugees is to let them stay in or near their own countries. This means providing assistance, aid, and, if we were honest, reparations, because we have a lot to do with the causes of the flight and migration.

During what was being called the American refugee crisis, starting in 2014, the largest group was people fleeing from Honduras.[7] Why Honduras? Well, it's a poor country with plenty of violence and destruction in general, but the violence dramatically escalated after 2009, when a military coup overthrew the parliamentary government. The United States was almost the only country that supported and legitimized the coup. It led to a sharp increase in killings and repression, and people started fleeing. When you do something like supporting a military coup, it has consequences. Just as there are consequences when you bomb and destroy Libya, or when you invade Iraq and smash it to pieces.

Africa is pretty much under the radar, sub-Saharan Africa in particular. Yet it's the site of enormous carnage and wars and destruction. Why don't we hear more about what's happening there?

There is a substantial U.S. military presence there, as has been exposed by the journalist Nick Turse, but it's under the radar, as you say.[8] The U.S. is conducting relatively small-scale military operations in Africa, and there aren't a lot of American troops involved, so we don't hear about it. In fact, we hear very little about any number of monstrosities. How much do we hear, for example, about eastern Congo, which is probably the worst disaster in the world? Millions of people have been killed.

The choice of what's reported or not has to do with special interests here, not with what's important.

In that regard, one of the biggest elephants in the room is Saudi Arabia, which doesn't really get enough scrutiny in terms of its actual policies.

Saudi Arabia is a violent and aggressive state. Its bombing of Yemen is exacerbating a very serious humanitarian crisis there. Not only do they bomb, but they bomb indiscriminately.

Hillary Clinton, not exactly a radical, said, "Donors in Saudi Arabia constitute the most significant source of funding to Sunni terrorist groups worldwide."[9] How does this feudal, homophobic, misogynist regime wind up as a major U.S. ally?

There's a three-letter word that explains it: "oil." They are the world's major oil producer. Also, they're obedient. Saudi Arabia is a family-run tyranny. Ever since a substantial amount of oil was discovered there in the 1930s, it's been a prime ally. In fact, during the Second World War, there was a conflict between Britain and the United States over who would control Saudi oil. Britain had been the major actor in the region before the war, but the United States pushed them aside and took over the huge Saudi oil concessions. Washington remains the dominant force in Saudi Arabia, sending them weapons worth tens of billions of dollars.

What Hillary Clinton said is correct. In fact, a European parliamentary commission drew essentially the same conclusion: Saudi funding is the main source of radical jihadi movements.[10] And Saudi Arabia is also the most extreme radical fundamentalist state. The British, when they ran the region, tended to support radical Islam rather than secular nationalism, and the Americans, when they took over, followed the same pattern.[11] It makes sense. Radical Islam has been a much more natural ally than secular nationalism. Secular nationalism carries the threat that governments might try to use resources for their own populations. Radical Islam has its own fanaticism, but it's not intrinsically opposed to imperial domination. In fact, it often relies on it.

The U.S. special relationship with Israel, which is unique in international affairs, is relevant here. U.S. relations with Israel were always reasonably close but only went completely off the charts in 1967. That's the year when Israel performed a huge service to the United States and Saudi Arabia. There was a major conflict under way then, a war between Saudi Arabia and Egypt. They were fighting each other in Yemen at the time, but the conflict was also much broader: Who is going to be the dominant force in the Arab Muslim world? Egypt was the center of secular nationalism in the Arab world, while Saudi Arabia was the center of radical fundamentalist Islam.

Israel settled that question: they smashed up the secular nationalist states, Egypt and Syria, and destroyed secular nationalism. I'm not saying that those were particularly attractive governments, but they were run by secular nationalists. And it was right at that point that U.S. relations with Israel changed radically.

The Iran nuclear deal has been described as "a stinging defeat" for the American Israel Public Affairs Committee (AIPAC), the major pro-Israel lobby in Washington.[12] Do you see the Iran deal that way?

This is a slightly unusual case. It's not just AIPAC that was against the deal. Strikingly, 100 percent of the Republicans were also opposed and voted against it. That's the kind of dedication to the party line that you do not typically find in political parties, with one exception: the old Communist Party. There, everyone had to follow the same line. That's one indication of how the Republicans have ceased to be a political party in the normal sense.

And why were they against the Iran deal? To some extent, they were just enacting the fundamental principle of the Republican Party ever since Obama got elected: destroy Obama and anything that might be seen as an accomplishment of the Obama administration. If he hadn't made the Iran deal, they would probably be in favor of it.

Their opposition also has to do with the way the Republican base has developed. As we've discussed, they can't get votes based on their actual policies, which are dedicated to the interests of the very wealthy and the corporate sector. So they've mobilized evangelical Christians and extreme nativists, and the people who have been harmed by the neoliberal policies of the past generation. After all, real wages for male workers are back to the level they were in the 1960s.

Median household wealth has actually declined in recent years.[13] There are plenty of angry, frustrated people.

This base is easily mobilized, especially the religious component. The evangelicals are probably the majority, or close to the majority, of the base of the Republican Party right now. For them, defense of Israel against Muslim attackers is a point of religious doctrine. After all, the Bible tells them so. They have a whole eschatology about it.

It's that group that was defeated by the Iran deal, but only temporarily. The Republicans actually had the majority in Congress, a substantial majority. Obama was only able to push the Iran deal through because their majority was not veto-proof. There's no question that the Republicans will continue to try to undermine the terms of the deal. And they may succeed in carrying out measures—increased sanctions, maybe secondary sanctions on other countries—that will lead Iran to withdraw from its arrangement with the United States. It's possible.

That does not mean, need not mean, ending the deal. Remember, this was not a deal between Iran and the United States, but between Iran and the so-called P5+1, the five permanent members of the Security Council and Germany. France, for instance, has established an agricultural trading mission in Iran.[14] The French are joining China and India, which for years found various ways to get around the U.S. sanctions, using barter instead of finance and the like.

The world is almost totally opposed to the U.S. position that Iran cannot have any nuclear energy program. The non-aligned countries have been vigorously supporting Iran's nuclear programs from the beginning. They're discounted in the West, but they actually represent the majority of the world's population. The U.S. may end up being totally isolated on this point, which would not be unusual—it's the same on many other issues.

Millions of dollars were spent on a campaign to oppose the deal, including full-page ads in major newspapers, TV ads. But they weren't successful.

They were successful in persuading a substantial majority of Congress. They were also successful in changing public opinion. If you take a look at the polls, at first the public was in favor of the deal. Over the months, as the propaganda campaign went on, support declined. The last polling I saw showed opinion is split fifty-fifty or even a little bit against.[15] So the opponents did succeed in gaining public support for their position, as well as congressional support. They didn't manage to override a veto, but they're at the point where they could definitely undermine the deal using the kinds of measures—like sanctions—they're now undertaking. And they're pretty open about it. They've announced what they're going to do.

If you go through the torture of listening to the Republican primaries, the debate is: Do we bomb Iran as soon as I take office, or—the moderate position—do we wait until after the first cabinet meeting, then bomb Iran?[16] To say that AIPAC and that whole amalgam—it's not just AIPAC—didn't succeed is a little misleading.

It's often said in the wake of this Iran deal that Tel Aviv–Washington relations have never been worse, that there's a major schism now dividing Israel and the United States. Does that hold any water?

Very little. Actually, Obama is probably the most pro-Israel president yet, though not pro-Israel enough for the extremists. This was obvious even before his first election, as I noted in 2008, just citing his website.[17] He had a very thin record, but one of the few things he did as a senator, which he advertised as one of his real achievements, had to do with the Israeli invasion of Lebanon. He cosponsored a resolution demanding that the United States do nothing to impede Israel's attack on Lebanon and, furthermore, that it punish anyone who opposed it. That's pretty extreme. It was a vicious invasion.

And he continued on that path as president. So, for example, in February 2011, Obama vetoed a resolution calling for implementation of official U.S. policy, which is that Israel shouldn't expand the settlements.[18] Of course, the expansion is a minor point; the real issue is the settlements themselves. This resolution called for stopping expansion and also noted that the settlements are illegal, which everyone recognizes. And Obama vetoed it.

Something even more important happened in the summer of 2015, and it barely got mentioned. Every five years the Non-Proliferation Treaty (NPT) participants have a review meeting. The Non-Proliferation Treaty's continuation is meant to be conditional on moving toward establishing a nuclear-weapons-free zone in the Middle East. That was an initiative of the Arab states, which have been pressing very hard for a weapons-of-mass-destruction-free region in the area.

Israel, which has nuclear weapons, is not a signatory to the NPT.

Israel, Pakistan, and India—all nuclear weapons states supported by the United States—are not signatories. Every five years, this comes up at the NPT meeting. In 2005, the Bush administration just didn't participate. In 2010, Obama blocked any discussion of a nuclear-weapons-free Middle East, and he did so again in 2015.[19] The U.S. gives one or another pretext, but everyone understands that the real motive is to keep Israeli nuclear weapons from being inspected and supervised. That's pretty serious. Not only does it create serious instability in the region, but it may also destroy the Non-Proliferation Treaty.

So Obama's position represents very strong support of Israel's military domination of the region. As Israel has moved far to the right, some of his views are now seen as hostile. But that's really more a comment on what's happened in Israel.

Let's go on to other subjects. What do you think about the partial decriminalization of marijuana in Colorado, Washington, Oregon, and California?

It's long overdue. Criminalization of drugs has been a social disaster. It's the main factor leading to the huge increase in incarceration. The United States is way ahead of everybody else in the world in tossing people in jail. It's a deeply racist system, as is evident in everything from police activities to sentencing practices. And it's deeply damaging, even after people are released from prison. People who have been convicted of drug possession, which is a nonviolent crime, can't get into public housing, can't get jobs, and so on. The only sensible thing to do is decriminalization, at least of soft drugs.

Think about tobacco, which is more lethal than marijuana, more lethal even than hard drugs. Tobacco use has declined along class lines, so it's now pretty much a class issue. Educated people with some degree of privilege are much less likely to smoke than they were twenty or thirty years ago. Tobacco wasn't criminalized. Instead, educational processes led to healthier lifestyles, better diets, and so on—and reduction of tobacco use was part of that.

Would you put alcohol in the same category?

Alcohol is also pretty much a class issue. It is far more lethal than drugs. Furthermore, alcohol and tobacco are not just extremely harmful to the user, but to nonusers as well. If you use marijuana, you're not harming anyone else. If you drink alcohol, you may become abusive and violent. There are many deaths of nonusers from alcohol—driving accidents, homicides, and so on. Yet, again, alcohol isn't criminalized; its usage has been controlled somewhat by educational processes.

Pope Francis's encyclical on the environment, Laudato Si', *generated some attention. He wrote, "Climate change is a global problem with grave implications," and he warns of an "unprecedented destruction of ecosystems, with serious consequences for all of us."[20] There is a movement to divest from fossil fuel. Are those divestment actions coming in time?*

Those actions are important, but they're nowhere near what needs to be done. The threat is far greater than reported. The scientific literature describes a pace of destruction that is already frightening, and that might at any moment become nonlinear, abruptly rising far more sharply. Even without that, even with just the regular processes that are predicted, there is likely to be a rise in sea level in the not very distant future. This could be massively destructive to countries like Bangladesh, with its coastal plains, and cities like Boston, a good part of which could wind up under water.

The Guardian *reports that "ExxonMobil, the world's biggest oil company, knew as early as 1981 of climate change—seven years before it became a public issue. . . . Despite this the firm spent millions" in subsequent decades "to promote climate denial."*[21]

That's what you expect to happen in a market society. Corporations are not benevolent institutions. They can't be. If they were, they wouldn't survive. They're dedicated to profit and market control. The same with health care: if you put it in the hands of private companies, they're going to try to make money from it, not concern themselves with health.

Which reminds me of a placard I saw in Seattle during the "kayaktivist" protest.[22] *The protesters were trying to block a Shell rig that was going to the Arctic to drill for oil. They held up a placard that said, "A good planet is hard to find." In terms of climate change, what can individuals do beyond, say, recycling?*

Recycling is worth doing, partly for itself, partly for symbolic reasons. It's a little bit like civil disobedience: the act itself may not achieve any end, but it does encourage others to do more. Eventually, though, we have to go beyond individual action to collective action. In our world that means actions by states, which have to be forced to take those actions by their populations.

Do you see the United Nations as a body that can bring about this kind of change?

No, the U.N. can act only as far as the major powers permit. It is not an independent agency. So when we ask, "Can the U.N. do something?" we're really asking, "Will the United States permit something to be done?"

Volkswagen has admitted that millions of its cars used software to defeat emissions tests. GM cars were found to have faulty ignition switches, which the company knew about and covered up, leading to more than 120 deaths. Laura Christian, the mother of one of these victims, a sixteen-year-old girl, said, "While nothing can bring my daughter back, we need a system where auto executives are account-able to the public and not just corporate profits."[23]

And not just auto executives. Johnson & Johnson, the huge pharmaceutical firm, is apparently facing billions of dollars of fines for mislabeling prescriptions.[24] Financial institutions are paying billions of dollars in fines for robbing the public. But that's the nature of capitalism: you try to steal as much as you can. We're after huge profits, and one of the ways to make prof-its is to cheat. And if you get caught and pay a fine, well, that's just the cost of doing business.

They pay the fines, but no one does any jail time.

As neoliberal attitudes and policies expand, there's more impu-nity. That hasn't always been true. If you go back to the savings and loan scandal under Reagan, quite a lot of people went to jail. That wasn't that long ago.

Bernie Sanders has been talking about income inequality and the depredations of the economic system, much as you've described. What do you think of his prospects?

It's interesting to see how much public support Sanders is getting with very little funding. He obviously doesn't have Sheldon Adelson giving him a billion dollars.

I think you can raise questions about many of his policies, but I think he's bringing important issues to the attention of a large part of the public. He's probably pressing the mainstream Democrats a little bit toward progressive directions.

But his prospects are pretty limited in a system like ours. It's a system of bought elections and what amounts to a plutocracy. And Sanders has a very small chance of breaking through this. But even if he did, by some near miracle, there isn't a lot that he could do. He's not running a political organization. He wouldn't have congressional representatives. He wouldn't have the bureaucracy. He wouldn't have governors, state legislatures, and so on. All the things that contribute to formulating policy would be lacking.

Concentrated private power is so enormous that it could block even somebody who had all of those supporting systems in place. So the chance that Sanders could implement any major policies would be slight, unless a massive political movement was behind him.

The real hope of the Sanders campaign is that after the

primaries—I assume he won't be nominated—the popular movement that supports him will persist, grow, and develop. That would be significant.

On the walk over to your office, I was talking to another MIT professor. I asked her, "If you could ask Noam Chomsky one question, what would that be?" She said, "Ask him, how does he do it?"

We are very privileged people, professors at MIT or elsewhere. We're reasonably well-off. We have a reasonable degree of security. We have resources, training. We're in one of the very few professions where you control your own work to a large extent. You may decide to work seventy hours a week, but it's your seventy hours, for the most part. There are commitments, but a lot of the work is work of your own choosing. That's very unusual in the world. Yes, there are problems and obstacles, and you can complain about this and that, but the opportunities are just enormous compared with what most people have.

You recently went back to your hometown, North Philadelphia. What was that like for you?

Actually, my wife, Valéria, wanted to see where I grew up and what it was like. It hasn't changed that much.

Were your parents very strict with you?

Strict? Only with the things they cared about. My father, for example, would insist that we have the right table manners. We could only get two ice cream cones a week. We went to Hebrew school, synagogue, and so on. I wouldn't say strict particularly, but there were rules.

How did you get along with your younger brother, David?

My skepticism about the adult world and recognition of its irrationality developed when my brother was just a couple of months old. My mother had told me this story about how it's going to be so much fun to have a baby brother, how I would have a playmate. And then this blob appeared, which did nothing but cry, get in my way, and take my mother's attention. They kicked me out of my room, so I had to sleep on a couch in my father's study. I didn't see any point to it at all.

One day, we went to the boardwalk in Atlantic City, and there was an organ grinder with a monkey. The monkey was fantastic. It was doing all kinds of wonderful things. I turned to my mother and asked her, "Why we don't trade in my brother for the monkey?" She didn't give me any reasonable answer, just laughed. And then I realized how ridiculous the adult world is, because it would have been an obvious trade.

But, later, we did get along like siblings, playing and so on.

Years ago, you told me that you had bad genes and didn't expect to live a long life. You're turning eighty-seven in December. How did you manage to trick nature, as it were?

I didn't do anything special. I didn't exercise or do all the things you're supposed to do.

As one ages, obviously there are infirmities and limitations. How are you managing that and still keeping up your work?

Some minor infirmities, naturally, but they've faded into the background since I met and married Valéria, which renewed my life.

You're inevitably asked in interviews, What gives you hope?

People who are dedicated, who are struggling, often against really tremendous odds—not like us—to create decent spaces for existence and a better world. That's my source of hope.

How important is solidarity and cooperation?

Without it, there is nothing. Individually, in an atomized society, you can do virtually nothing. You can ride a bicycle instead of driving, say, but that's kind of like chipping away at a mountain with a toothpick. If anything is going to happen, it is going to be through mutual aid, solidarity, community, and a collective commitment to really making changes. That's always been the case in the past, and there's no reason to think it will be different in the future.

8

THE ROOTS OF CONFLICTS

Alexander Cockburn used to quip that the two greatest disasters to befall the United States in the twentieth century both happened to occur on December 7th: the Japanese attack on Pearl Harbor and your birth in Philadelphia.

I can't deny it. It's right there in the hospital records, so it must have happened.

Well, happy birthday. Eighty-seven years young.

Did I ever tell you, my name is actually wrong on my birth certificate? I once had to look up my birth certificate for some reason, so they sent me a copy from city hall. At the time, apparently, the clerk didn't believe that my name was Avram Noam. He thought Noam must be Naomi, and wrote that in. Then Avram had to be a girl's name, so he made a handwritten correction there too: "Avrane." So my birth certificate reads, "Avrane Naomi Chomsky."

In 1966, you gave a talk at Harvard that was published the following year in the New York Review of Books. *It was your famous talk "The Responsibility of Intellectuals," which put you in the public spotlight in terms of your political work.[1] You were already well established in linguistics.*

I had written political articles before, but that was the first one to appear in a major journal and be read by more than a few activists.

Regarding the responsibility of intellectuals, what can you say about the current generation? Are they any different?

I don't think it's been different for all of recorded history. The phrase "the responsibility of intellectuals" is actually ambiguous—and intended to be ambiguous. There's a responsibility that they're expected to fulfill, essentially to be flatterers at the court. Then there's the moral responsibility of being truthful, accurate, critical, focusing on crimes for which we share responsibility as part of our state, our society, whatever it may be.

About U.S. motives globally, you say "a useful way to approach the question . . . is to read the professional literature on international relations" in order to understand "what policy is not."[2]

Well, there are basically two theories of international relations. One of them is called Wilsonian idealism; the other is called realism. If you look at them closely, "realism" is not particularly realistic. In fact, it tends to ignore such crucial factors as the determinants of power and decision-making in the domestic system. Wilsonian idealism, meanwhile, simply repeats the basic illusions of every imperial power: that we are exceptional, that we may make mistakes but we always have good intentions, and so on. That's precisely what policy is not.

You said in a recent interview that U.S. policies have "succeeded in spreading jihadi terror from a small tribal area in Afghanistan to virtually the whole world, from West Africa through the Levant and on to Southeast Asia."[3] How did they do that?

When the only method you have is to use your comparative advantage in violence, you will always make the situation worse. The military analyst Andrew Cockburn points out that every time you kill a leader, you think it's a big triumph.[4] But what you're doing, almost invariably, is replacing him with a younger, more competent, more violent leader. It happens over and over.

In fact, we're doing jihadis a favor. ISIS and Al Qaeda, for example, fairly openly said, "Please come attack us. Send the crusader armies to fight us. It will be a recruiting tool. Pretty soon you will be at war with the whole Muslim world." That's just what they want.

Let's say ISIS is bombed to smithereens. What then? Do the troops come home? Does the U.S. close down its overseas bases?

No, because ISIS would be replaced by something worse. I don't know exactly what, but there are other groups. Like it or not, in

much of the Sunni world, ISIS is regarded as providing a kind of
protection and security. And apparently they run, by some
standards, a fairly efficient totalitarian system. It's kind of like
Iraq under Saddam Hussein. He was a brutal dictator, but
people had security, they had education. As long as you shut
up and didn't talk about politics, you could have a pretty
decent life—in fact, more so than almost anywhere else in the
Arab world. Now they have nothing, just war. If we don't deal
with the roots of the problem, then something worse will arise
from the same causes.

*Do the policy makers and managers of the U.S. state system con-
sciously promote conflict and turmoil?*

No. Take, say, Libya. Muammar Qaddafi was a brutal, ugly
guy. But he managed to build some kind of functional country out of a tribal society. There was an uprising, and he put it
down quite harshly. I think about a thousand people were
killed. Then three imperial powers—France, the United States,
and Britain—rammed through a Security Council resolution
that called for a cease-fire, protection of civilians, negotiations,
and diplomacy.[5] Okay, what did the outside powers do?

They violated it.

Qaddafi accepted the cease-fire, but the imperial powers
instantly violated it and became the air force of the rebels. The
result was to destroy the country, to radically magnify the number of casualties, and, incidentally, to spur the flow of refugees

from Africa to Europe. Did they plan that? No. It's just the famous adage: "If all you have is a hammer, everything looks like a nail." We have a hammer. We're really good at smashing things.

Some people say, "Well, look at the benefits that accrue to the weapons manufacturers here. Isn't that the rationale for more bases, more intervention?"

That's a factor, but I don't think it's the driving factor. The driving factor is that the traditional role of any great power is to expand its power. England, France, and the United States each have a long history of imperial domination. So they do what comes naturally. They use their comparative advantage, which happens to be not diplomacy, development, freedom, or anything like that. That's what they talk about, but what they're really good at is force. So we have the Joint Special Operations Command (JSOC), Special Forces, drones, armies that can go in and smash places up.

In case after case there have been diplomatic alternatives. Whether they would have worked, we can't say.

What about alternatives for alleviating the misery and suffering in Syria?

The situation in Syria is terrible. There's one region in Syria that is doing all right, the Kurdish region of Rojava. The Kurds have succeeded in defending it with very limited weapons, and have apparently managed to create a pretty decent, functioning, interesting society under horrible conditions. Aside from that, the place is just a wreck, nothing but gangsters murdering each other. The Assad regime is brutal and destructive, and its atrocities have caused most of the deaths.[6] Then there's ISIS. There's also a major jihadi group, an offshoot of Al Qaeda, the Al Nusra Front, and another one that's not very different, Ahrar al-Sham. The two of them sort of merge.

Did you hear Prime Minister David Cameron calling for the bombing of Syria in the British Parliament? He said, We have to support seventy thousand democratic freedom fighters over there. Robert Fisk asked in an article the next day, did he mean seventy thousand or just seventy?[7] Every correspondent who knows anything about Syria has ridiculed Cameron's claim. Nobody can find these people.

What's the answer? There's just one option, even if it's only got a slim chance of working, and that is some kind of negotiations involving every warring group in Syria—apart from ISIS, which has no interest in negotiations. This means arranging negotiations among real monsters. You don't like any of them, but that's the only choice. If you want to reduce the killing and the destruction, that's what you do. Maybe they can work out local cease-fire agreements that would reduce the violence, make plans for some sort of transitional government, ultimately maybe elections. Again, this is separate from the question of the Kurdish areas, which should simply be protected in whatever way we can.

The U.S. has been blocking this strategy until recently on the grounds that we can't allow a monster like Bashar al-Assad to participate. But that's tantamount to saying, "Let's let them murder each other." Whatever you think about Assad, he's not going to commit suicide. So if you want the problem to be solved without total destruction, he will have to be part of the negotiations.

But what is being done? None of the above. Syria is in danger of being finished off as a viable country. The idea that you can find a military solution is ludicrous.

What prompted Putin's intervention in Syria?

We don't know in detail, but it appears that what prompted it in large part was the CIA supplying heavy weapons, including wire-guided anti-tank missiles, to the jihadi elements. These were taking a huge toll on the Syrian army. Russia supports Assad, so apparently they moved in substantially in response to that. We can't prove this, but that's what the limited information we have appears to suggest.

You recently spoke via Skype to, I believe, an Israeli university, and you managed to upset some people. What was their objection?

Actually, it was the Van Leer Jerusalem Institute, which is a research institute near the Hebrew University. The talk was a memorial for an old friend of mine, Yehoshua Bar-Hillel, an Israeli philosopher and logician who died some decades ago. I had known him for years. His daughter Maya, who teaches at the university, organized a memorial for the hundredth anniversary of his birth. I was quite close to him, and knew his wife and the children when they were younger.

A number of Palestinian intellectuals, some of them friends, called on me to cancel the talk, because they believe that we should boycott all Israeli universities on principle. I don't agree with this, just as I didn't agree with such a policy in the case of South Africa. If you have a targeted boycott—say, against racist hiring practices, in the case of South Africa—it makes sense. But when Howard Zinn went to talk at Cape Town University, for instance, I didn't see anything wrong with that. I thought it was good for the South Africans, and they thought so, too. In any event, I decided I would go ahead and give the talk (by video-conference) for the Van Leer memorial. So, yes, that outraged some people.

It's kind of ironic. The last time I went to Palestine to talk at a Palestinian university, I was blocked by Israel. This time I was protested by Palestinian intellectuals.

What's the current state of U.S. imperialism? In scholarship and in the corporate media, one can rarely use that term. Where do you see the empire today?

First of all, we should say that with respect to scholarship, the situation is changing. The main scholarly journal of diplomacy, *Diplomatic History*, had a very interesting article tracing U.S. imperialism back to the first colonists, pointing out that the conquest of the continent was in fact imperial conquest.[8] That's the way it was viewed by the Founding Fathers. Later, that view disappeared. The article castigates historians for treating U.S. history as if imperialism began in 1898.

So scholarship is beginning to come to terms with the history of imperialism, and even drawing self-critical attention to the fact that we practically exterminated the indigenous population of the continent. We violated every imaginable treaty with them, and so on. This is what the historian Richard Van Alstyne called the "rising American empire," and it began with the colonists.[9]

Imperialism basically means domination of others, and it takes many different forms. It can take the form of overt rule over the natives. It can take the form of settler colonialism, the worst kind, where you drive out the natives and replace them. There are other forms, too, such as economic domination. Take the so-called free-trade agreements, like the North American Free Trade Agreement, the Uruguay Round of the World Trade Organization, the proposed Trans-Pacific Partnership. These have nothing much to do with trade, despite the name; they are largely investor-rights agreements. They give multinational corporations and investors substantial control over the resources, policies, and actions of other countries. You can call that imperial domination, if you like, or you can call it something else. These are not well-defined terms.

Do you see the U.S. position today as stronger or weaker than it used to be?

It's weaker.

Why do you say that?

For one thing, U.S. power has been declining for seventy years. At the end of World War II, that power was phenomenal. The

United States had perhaps half the world's wealth and total security. Every other industrial society was devastated. That couldn't last, obviously. As the other industrial societies rebuilt, and decolonization pursued its anguished course, gradually power in the world was to some extent distributed. But though it has declined, the United States remains overwhelmingly more powerful than other countries.

Particularly in the military sector. Less so in the economic sphere.

The military sector is unmatched. Economically, the situation is more complicated, but the United States is still in a dominant position. Europe, the other major economy, is powerful, but it has decided to undermine itself through self-destructive economic policies. Austerity during recession has been very harmful.

China is the biggest economy in the world, in terms of purchasing power. On the other hand, it's still a weak society, and its per capita income is quite low. If you look at the United Nations Human Development Index, which measures various properties of a society, China ranks ninetieth.[10] And the country has huge internal problems, which are going to increase in the coming years.

Labor issues, income inequality, environmental problems.

Yes. There's a lot of labor militancy, thousands of labor actions every year. There is also the demographic problem. There was a demographic peak of people in the roughly twenty-five to forty age range, which gives you a huge workforce—

but that's declining. Not as much as in Europe, but significantly.

China is basically a poor country with a big export economy, a lot of which is owned by outsiders. If Apple produces iPhones at Foxconn and exports them, those are called Chinese exports, but China doesn't get all that much from it.

Nevertheless, the International Monetary Fund has designated the Chinese renminbi as a global currency, joining the U.S. dollar, euro, British pound, and Japanese yen.[11]

China has plenty of capital. They've accumulated a lot of it from their exports. So in financial terms, they're quite well off, and they spend money at a level well beyond other countries. The Chinese have many investments in central Asia. They're slowly rebuilding something like the old Silk Roads, several of them, that run through central Asia, including to a big port in Pakistan they've developed.

Gwadar.

Yes. Chinese trade with the Middle East, if it goes by sea, has to go through areas which are largely controlled by the United States and its allies. So they've been quietly building alternatives. The port in Gwadar, if it works, will be connected to transport systems and pipelines that go straight to eastern China. That will be a way for, say, oil to flow to China.

China is also investing very heavily in Africa and other places. But while they are a huge power financially, I don't think they compete with the United States economically.

China's high-speed rail network, however, is a model.

Yes, though the United States is unique in how badly it handles rail. There's an ideological resistance here to effective public transportation, to the point that some Republican-run states even refuse free money from the federal government to build high-speed rail. It's ridiculous, and it harms the economy enormously. But it's very important to some people to ensure that we don't have public services that are efficient and effective. You want to maintain domination by private capital. It's the same reason we don't have a good health care system.

Japan is moving to amend its constitution to allow for the use of military force.[12] What's the significance of that?

It's very significant. Japan has a so-called peace constitution. Article 9, imposed under U.S. occupation, explicitly states that Japan has to abandon its imperial pretensions and to commit whatever military forces it has solely to self-defense, with no participation in foreign military actions. If you look at the 1930s and 1940s, you understand the reasons. But there has been conflict over Article 9 in nationalist circles for some years. The government of Prime Minister Shinzō Abe has moved pretty vigorously to try to modify the constitution. They haven't rescinded Article 9, but they're reinterpreting it so that it permits Japan greater action in non-defensive extra-territorial actions.

Japan hosts a huge number of U.S. military bases, mostly in

Okinawa, which, though part of Japan, is virtually a colony. The Okinawans don't want the bases, but the Japanese government has overruled the Okinawan population, even their elected officials. The United States is now planning to build yet another base, which has provoked protests as well.[13]

From the Chinese point of view, of course, this is all aimed at China. Okinawa is no small thing. There is a history, which maybe we don't want to know about but the Chinese are quite aware of. In 1962, for example, six months before Nikita Khrushchev sent missiles to Cuba, Kennedy sent intercontinental missiles aimed at China to Okinawa. At the time, there was a small war going on between India and China. And China was also in conflict with Russia. It was a very tense moment. So Kennedy sent these missiles to Okinawa.[14] Of course, we don't talk about that. We only talk about Russian missiles in Cuba—not U.S. missiles in Okinawa, Turkey, and elsewhere.

China is ringed by offensive missiles in hostile states under U.S. domination. And there are more and more of them, including on Jeju Island, South Korea, which has an effective U.S. military base. It's part of the whole confrontation with China over the South China Sea. China itself is also taking aggressive measures, including building artificial islands, which interfere with the sovereignty claims of other countries.

Is the U.S. government re-creating the "containment" policy?

It's not re-creating it, because it never ended.

You project an image of tranquillity and serenity, but you once told me that your stomach is often churning when you're speaking. Where do you get your apparent equanimity from?

Maybe from my father, who was kind of a stoic. He always managed to keep a calm exterior. Maybe from him. Who knows?

Your mother, Elsie, was a teacher. And I read somewhere that she was politically active. Is that right?

She was active in Zionist, Hebraic, Jewish cultural circles. She was one of the leading intellectual figures in the Hadassah women's organization. That's where her political activity took place. Her family was also very active politically. They were working-class New Yorkers, mostly unemployed, involved in political factions, the Communist Party and others.

And that would be your mother's sister Sophie's husband, Milton Kraus?

Yes. Actually, Sophie and her husband met at a demonstration, I think. Sophie may have been close to the Communist Party— I'm not sure—but Milton wasn't. He had broken free from the factions and was way off to the left by himself.

He was legally blind, I believe?

He was. But he could get around and he could read, with effort. He read a lot, in fact.

And he ran that now legendary newsstand on Seventy-second and Broadway.

He was also physically deformed. He was hunchbacked and small. It was how he got the newsstand, under a New Deal disability provision.

As you got older and as your criticisms of Israel became more and more pronounced, did you have any constraints or feel any difficulty in talking about this with your parents?

They weren't happy about it. I had correspondence with my father about it, and we had discussions about it. Not so much with my mother, who didn't object, but my father. It's not that he disagreed. He said that he more or less agreed, but he didn't like the way I was talking about it. He didn't say it exactly, but there's an old Jewish custom—you don't talk about certain topics in front of the goyim.

Your book Peace in the Middle East? *was published in 1974, after your mother had passed away, but your father was still alive.[15] Did he read the book before it came out?*

I had sent him chapters. Some of the chapters were first written in the 1960s, so he had seen them.

In many families, the mother is more affectionate and the father is the austere figure. Did that play out in your family as well?

Not really. My father was mostly engaged in his work, but I had more personal relations with him than with my mother. When

I was nine or ten, we would spend a couple of hours on Friday evenings reading Hebrew articles, essays, poetry, and so on together.

Bob Teeters was a close childhood friend of yours. There's an interesting story about his mother and your father.

I went to school from about age two. My parents were both Hebrew teachers, so they worked in the afternoon. When I was a baby, there was somebody, a maid, who would pick me up. But by the time I was maybe five or six, so probably by first grade, I would start going over to Bob's house in the afternoon, and his mother would take care of us. Bob was in the same class, and he lived right across the street from the school. My parents would show up around six-thirty or so, after Hebrew school, and pick me up. That went on until we were in eighth grade. Then he went his way to high school, I went mine. We barely saw each other after that.

Our parents were friendly but not socially friendly. They didn't have dinner together or anything. But many years later, Bob's father died, my mother died, and somehow his mother and my father got together, and they finally got married. She was a midwestern Christian, and fairly anti-Semitic. She made an exception for my family, but in general she didn't like Jews. She did convert to Judaism, which she didn't like either, but my father insisted. It was complicated. But they were happy together.

Do you have any birthday celebrations planned?

That's personal. I don't talk about it.

People are interested in the man Chomsky.

They might be, but I've always kept that very separate from my public life.

9

TOWARD A BETTER SOCIETY

CAMBRIDGE, MASSACHUSETTS (MARCH 11, 2016)

"Socialism" was one of the most looked-up words in 2015, and several polls indicate that young people ages eighteen to twenty-nine have favorable views of socialism.[1] Are you surprised by that?

Not so much surprised as uncertain. The question is what they mean by it. I suspect they mean something like social democracy, which is essentially New Deal welfare-state capitalism. And if that's the case, it's not surprising, because polls have shown for years that this matches widely held goals of most of the population.

You've been involved in myriad struggles and actions over the decades. Are there any you feel have lessons for today?

Just about all of them. Take organizing poor people. The civil rights movement has plenty to teach about that, as does the labor movement. There are some common elements to all kinds of organizing; you have to find issues that meet several conditions. First, people have to care about them. Second, the fixes have to be feasible. And third, it has to be possible to convince people that they are feasible, because one of the major impediments to organizing is the feeling of "you can't fight city hall." So you have to show that you *can* fight city hall. The way it's typically done in successful organizing is to find small things that people recognize could be achieved, see if you can achieve those, then encourage the sense that success is possible, and proceed to the next thing.

Take a real case that I heard of not long ago, about a group working in working-class immigrant communities in South Boston. At the beginning, just to break through the sense of hopelessness, they started with something very simple: women organized to see if they could get the town to put in a traffic light where their kids cross the road to go to school. They worked on the issue, they pressured local officials, and they won. Then they recognized that they could do things if they worked together. And they went on to the next campaign. That's how you build. That's organizing and activism.

Let's talk about Latin America, a region historically under the U.S. thumb. Hugo Chávez was elected president of Venezuela in 1999. He was succeeded by Nicolás Maduro. What's your evaluation of what's happened there?

What's happened in Venezuela, first of all, is an extreme case of what's happening in Latin America generally—and it's pretty tragic. Chávez himself tried to create significant and positive changes, but the way he did it had fundamental flaws. For one thing, it was top-down. It was not coming from the base. There were some efforts to organize popular activism, but it's very hard to find out—at least I haven't been able to find out—how successful they were.

Then there was a significant amount of corruption and incompetence that seriously undermined his efforts. How high it went, we don't know. Finally, there was no real dent in the reliance of the economy on a single resource, oil. In fact, Venezuela probably became even more dependent on it. Venezuela could have a rich agricultural economy, a productive industrial economy, but instead its economy is overwhelmingly reliant on oil exports.

Actually, I think Chávez himself was aware of this. He gave an important talk at the United Nations where he pointed out that Venezuela is a fossil fuel exporter, but said that the producers and the consumers really ought to get together to work out ways of getting the world off fossil fuels, because they are so destructive. That's a very unusual position to take for someone whose economy relies on fossil fuel production. Chávez was widely ridiculed for calling George W. Bush "the devil" at that meeting, but I didn't see any reporting of his comments on fossil fuels.[2]

The combination of initiatives coming from above, the failure to move toward diversification, the corruption, the incompetence—all of those together have led to the collapse of the economy in Venezuela.

As for what's happening in Latin America generally, it's highly significant. Latin America is a potentially rich area, with countries that could be wealthy, advanced. Over a century ago,

Brazil was considered "the colossus of the South," parallel to the colossus of the North. These countries have rich resources, don't suffer from external threats; they have lots of opportunities for development.

But that development hasn't happened. The main reason is internal. The countries have typically been dominated by small, Europeanized, mostly white elites who are enormously wealthy and linked to the West culturally and economically. These elites do not assume responsibility for their own countries, which leads to horrible poverty and oppression. There have been efforts to break out of that pattern, but they have been crushed.

However, in roughly the last fifteen years, there have been some attempts to address these issues in several countries— Brazil, Venezuela, Bolivia, Ecuador, Uruguay, Argentina. It's been called the "pink tide," and it has met with varying success. There's just a very strong temptation, when you gain a bit of power, to put your hand in the till and live like the elites. This has undermined left governments in case after case. Venezuela is a case in point. Brazil is another. The Workers' Party had a real opportunity to change not only Brazil but all of Latin America. It did achieve some things, but it also substantially squandered the opportunity.

So my feeling is there are solid gains, and then there's regression, which eliminates some of those gains. There might be a basis for moving forward again in the future, if the current issues are resolved. But it's not certain that they will be.

Luiz Inácio Lula da Silva of the Workers' Party became president of Brazil in 2003. He was succeeded in 2011 by his protégée, Dilma Rousseff. Corruption scandals, particularly around Petrobras, the state-owned oil company, have sunk her approval ratings. Yet just a few years ago, Brazil was being touted as one of the so-called BRICS economies— Brazil, Russia, India, China, and South Africa—that were going to provide an alternative to U.S. domination.

Yes. In fact, Brazil was in many ways one of the most respected countries in the world. Lula himself was very highly respected— by me, too, I should say. Among world leaders, he's a pretty honorable one, I think. I'm surprised by the corruption charges and a little suspicious of them. I don't know to what extent this is some sort of right-wing coup and to what extent it's something real. The charges that have been made public aren't very convincing. So we'll wait and see what comes out. I don't think the facts are clear at this point. But it's true that the corruption was very serious.

There are parallels between Brazil and Venezuela. Brazil also did not make use of its opportunity to move toward a more diversified economy. In fact, it moved away from it. Brazil benefited temporarily from the very rapid growth of China, and its huge appetite for Brazilian raw materials such as soy and iron. Of course, relying on that has a consequence. It means importing cheap Chinese manufactured goods, which undermine your own manufacturing capacity. The end result was that the Brazilian economy remained resource-based instead of diversifying. This is also true of Argentina and Peru. It's not a viable mode of economic development.

After the American Revolution, if the United States had followed that path, we'd still be exporting agricultural goods, fish, and fur. In fact, that was the prescription given to us by the leading economists of the day, like Adam Smith, who made much the same arguments that the IMF and neoliberal economic development experts are making to the Third World today. The United States didn't follow the economists' advice, because it was independent.

In 2006, Evo Morales, in Bolivia, became the first indigenous person to be elected president of a Latin American country. In 2016, he was defeated in his bid for a fourth term as president.

Partly it's an anti-caudillist sentiment: let's not go back to the system of a powerful leader who remains in power forever. To Morales's credit, he accepted the defeat. I also think some of the same problems I've discussed, as well as achievements, exist in Bolivia as well.

Let's move on to India and the ruling Hindu nationalist party, the Bharatiya Janata Party (BJP), and its leader, Prime Minister Narendra Modi. The Guardian *reports, "The government has repeatedly been accused of seeking to repress free speech and encouraging extremist nationalists who systematically intimidate critics." Students at Jawaharlal Nehru University (JNU) have been arrested and accused of "anti-national" activities and sedition.[3] What's going on there?*

The students were calling for a free Kashmir, opposing the cruel and bitter Indian repression there. They also supported an activist who was accused of crimes and subsequently executed.

That was Afzal Guru.[4]

Yes. The students questioned the validity of the charges against him. The police were called in to repress the protests. They arrested a student leader and picked up faculty members for harsh interrogation.[5]

Similar things have been happening in other universities. And it's all taking place against a background of increasing Hindu nationalist violence and repression—murdering a Muslim because they claimed he stole a cow, for instance.[6] Things like that are going on around the country.

Hindu nationalism, like other forms of extremist nationalism, is a frightening phenomenon. It's another one of the moves toward authoritarian nationalism and religious extremism that we're seeing around the world.

Can you talk about India's growing military ties with the United States and Israel?

That's a real change. Under Jawaharlal Nehru, India was a core part of the nonaligned countries. Its military links were closer to Russia. In recent years, though, it has increasingly shifted to become part of the U.S. orbit. That includes closer relations with Israel.

A central cause of this shift is the anti-Muslim sentiments common to all three countries: Israel, obviously—where not just any Muslim but any Arab is seen as a threat; India, which has a big Muslim population in a Hindu-dominated state; and the United States, where you have extensive anti-Islamic sentiment among the general population and a so-called war on terror directed against Muslims.

The United States has actually supported the development of nuclear weapons in India, as well as in Israel and Pakistan. Those are the three countries that have not signed the Non-

Proliferation Treaty. According to the terms of the treaty, the United States was not supposed to provide any nuclear assistance to countries that hadn't signed the NPT; but under the second President Bush, the U.S. went ahead with providing aid to nuclear development in India anyway.[7] The U.S. government claims to only support India's civilian nuclear facilities, but that's meaningless because, first of all, the aid is transferable from civilian to military uses, and secondly, the aid frees up India to devote more of its own efforts toward nuclear weapons. The Bush administration also succeeded in twisting the arms of other countries, so the Nuclear Suppliers Group went along with essentially building up India's nuclear weapons power.

By now the U.S.-Indian nuclear alliance is quite tight. At the same time, the United States is also supporting India's main enemy, Pakistan, as is China. China and Pakistan have a common interest in what's called anti-terrorism. In the Muslim regions in the western part of China, Uighur areas, there are guerrilla and other actions that China is repressing pretty harshly. Those groups have links to the Pakistani Taliban. So there's some cooperation between the two governments.

There is a possibility that Chinese aid to Pakistan will actually become developmental in character, which could be positive—in contrast to U.S. aid, which has been overwhelmingly military. The infrastructure that China is slowly constructing throughout Eurasia, which pretty soon will reach all the way to Europe, is a major development in world affairs. Control over Eurasia has been recognized for a long time to be critical to global power, sometimes even regarded as the key to global power.

Turkey is another country that has been sliding toward more and more autocratic and authoritarian rule. Newspaper offices are raided, journalists and academics are threatened and arrested. Erdoğan, the president of the country, mentioned you recently by name and in fact invited you to visit Turkey.[8] What is the background of that? And did you make that trip?

I didn't. As for the background: there was a terrorist bombing in Ankara that killed a lot of people. It was probably ISIS—which Turkey is tacitly supporting, incidentally, in many ways—but Erdoğan blamed it on the Kurds.[9] That led to significant repression against the Kurds. There have now been several months of intensive curfews in southeastern Turkey, affecting hundreds of thousands of people. The curfews are very harsh and brutal.[10] People can't leave their houses. There are snipers on the rooftops, heavy military equipment. There have been many deaths. Sometimes dead bodies are left in houses to decay because people can't remove them.

Turkey provides a kind of a funnel through which jihadi fighters can travel to the so-called Islamic State territories in Syria. It's exporting oil from ISIS. Some claim it even has hospitals on the Turkish side for ISIS fighters. I don't know if this is true, but the journalists who reported this were immediately jailed.[11] That led to a protest, with a petition against the jailing signed by a thousand or so Turkish academics.[12] I was one of the international signatories. And when Erdoğan reacted by attacking the academics, which could have serious consequences for them, he also castigated me personally for this terrible attack on Turkish honor. He said, Why don't you come here and see what the reality is? I was asked to write a comment, so I wrote a couple of lines.[13] But that was the end of that.

In northern Syria, a mostly Kurdish area called Rojava is apparently inspired by the work of Murray Bookchin, a U.S. writer and thinker who died in 2006.[14] What's going on there?

It's not entirely clear. For one thing, the circumstances are horrible. Syria is just imploding, and the conditions are awful. There's constant fighting, though a semi-truce exists between the Kurds and the Assad government, which don't seem to be attacking each other much lately.

The Kurds are perhaps the main ground force defending the population against the ISIS monstrosity. Their leader is Abdullah Öcalan. He came from a strong Stalinist background, but during his time in prison, he shifted his attitudes considerably—at least in his writings—and picked up from Bookchin a kind of communitarian, anarchist approach. Kurdish groups, both the PKK in Turkey and northern Iraq and the Kurds in the Syrian Kurdish area of Rojava, appear to have been influenced by these ideas. To what degree the ideas have actually been implemented is hard to say, though everyone agrees there has been a new emphasis on women's rights and women's involvement, as well as some communal programs. It's pretty remarkable under the circumstances.

Elections are once again front and center. Why aren't elections in the United States held on the weekend, when people are not working, as in other countries? What about abolishing the electoral college, which seems to be antediluvian? And what about allowing people to vote where they are currently living rather than where they first registered?

There are indeed major differences between elections in the United States and in other capitalist democracies. For one thing, elections here never stop. As soon as one election is over, you start working on the next one. That means the day you walk into your office, you start fund-raising for the next campaign. Of course, that affects policy decisions. In other countries, there's just a brief period of campaigning, some debate, then people vote.

The role of money in U.S. politics also goes way beyond other Western countries. It's been significantly increased by a number of Supreme Court decisions, going back to *Buckley v. Valeo* on the money-is-speech issue, then the *Citizens United* decision and others. But it goes way back. There's pretty convincing evidence that campaign funding is a good predictor of policy decisions.[15]

Then there's radical gerrymandering—mostly by Republicans, but others, too—as well as major efforts by the Republicans to keep people from voting. There is good reason for them to do this: the people who are most vulnerable are less likely to vote for the Republicans. So the Republicans don't want Sunday voting, when black churches might take people to the voting booths. They want to require the kind of identification for voting that many poor people, and particularly black people, just don't have. These are all efforts to maintain power despite minority support.

The most striking example, perhaps, is the House of Representatives, which is almost a Republican lock, even though they get a minority of the votes. In the 2014 election, the popular House vote was in favor of the Democrats, but the Republicans won the majority of seats.[16]

They won the House because of gerrymandering.

Not just gerrymandering. Demographics also play a role: Democratic voters happen to be primarily urban, while the Republicans have a scattered rural vote. That means that you get a heavily Democratic vote in an area where there are just a few representatives. But that's not to deny the importance of gerrymandering. And since the Supreme Court threw out the Voting Rights Act, which had protected minority voters in states with a long history of racism and voter suppression, it's gotten much worse.[17] All these things together, I think, are far more significant than, say, the electoral college. Of course, there's plenty wrong with that institution, but it's marginal compared with these other factors.

The United States is unusual in that it doesn't have class-based political parties. It has geographically based parties, and rather odd coalitions. In fact, in many respects, the party system still reflects the Civil War. Take a look at the red and blue states in the 2012 election. It's almost the Confederacy and the Union, just by different names. A recent study published by Brandeis University actually found a pretty clear correlation between Ku Klux Klan activity not that many years ago and current Republican votes.[18]

Also, abstention is very high in the United States. Walter Dean Burnham, one of the leading specialists on politics, did an interesting study in the 1960s on who doesn't vote here.[19] It turns out that going by their socioeconomic profile, non-voters in the United States are quite similar to voters in European countries who vote for social-democratic or labor-based parties. Those parties simply don't exist here, so comparable people don't vote.

So you have a variety of factors. Lack of proportional representation has a big effect; because of the winner-take-all system, you can't develop independent parties here. There's no chance for them to grow slowly and perhaps become dominant over time.

But there are positive sides to our political system, too. For example, freedom of speech is much better protected here than elsewhere. There are serious problems, but it's not an all-or-nothing story.

You might have heard the joke that if God had meant us to vote, he would have given us candidates.

I understand it, but it's a bit too cynical. There are differences between the candidates, sometimes significant ones. Incidentally, it's not all about the party labels. I have voted for Republicans myself, for example. During the 1960s, the Republicans were the ones who were more strongly anti-war in the state elections.

But in recent years, the Democratic Party has been a mildly centrist party, while the Republicans have shifted very far to the right. And they have very significant differences on some issues, though the two biggest ones—life-threatening, species-threatening issues—are barely discussed in the elections by either side. One is global warming; the other is militarization.

On those topics, there's a major difference between the two parties.

In a discussion you had at the Harvard Trade Union Program with activists from around the world, someone asked you about hopeful signs for the future.[20] I was taken aback by your answer. You mentioned Egypt, citing a book by Jack Shenker, a former Guardian *correspondent there.[21] Egypt is going through hell right now.*

It is. And when I read Shenker's book, I didn't really trust my own judgment. I don't know that much about Egypt. But I sent the book to friends, some of whom really do know Egypt very well and are very good analysts, and they thought he was pretty accurate. What he found is that although the Sisi dictatorship is driving the country to disaster, nevertheless a good deal remains of the vital activism and achievements of what's called the Arab Spring—particularly in the labor movement, which he's looked at quite closely. I think that's an interesting sign.

The *New York Times* has a story about the laments of the Egyptian elite, the rich students, people who have benefited from the dictatorship.[22] They're very upset. They have to wait a month to get a Mercedes, and they may not get the kind they want. In general, the benefits of standing with those in power are apparently getting harder to acquire, which is leading to dissent among the elite supporters of the regime. And the regime is in economic trouble. Saudi Arabia is no longer funding the Sisi regime at anything like the level they were.

A journalist friend of mine in Gaza calls me fairly regularly on the phone. He lives near the Rafah border, the Egyptian border, and you can hear the shelling. He says he hears it all the time. The Egyptians have not been able to suppress the Bedouin uprising in the Sinai, which is probably a pretty big drain, too. The downing of that Russian airliner, killing a couple hundred people, also had a major impact on the tourist industry, which they rely on.[23]

So the regime is in some difficulty. And if Shenker is correct, which he may well be, there are still the germs of what could grow into another one of Egypt's many efforts over the years to create a more democratic society.

Among the questions you pose in your book What Kind of Creatures Are We?, *I want to ask you about two: What would a decent society look like? And what would satisfy our basic needs and rights?*[24]

I'm not smart enough—and I don't think anyone is smart enough—to sketch what an ideal society would be, but I think we can discuss what would be a much better society. To my mind, it would be primarily a society in which decisions are in the hands of an informed and engaged public. That's a prerequisite for being reasonable and rational in your choices. As for institutions, it would mean that workers would own and run factories, communities would be under community control, other institutions would be under popular control. Interactions among voluntary associations would lead to broader decision-making, all by representatives who are under direct control from below and subject to immediate recall.

There would also be a fading away of national boundaries, which is certainly conceivable—it's already taken place to an extent in Europe. In general, it would mean an increasingly global system based on mutual aid, mutual support, production for use rather than profit, and concern for species survival. Those are all directions toward a better society, I think. And they are all feasible.

You said to me recently, "There's a lot of dry kindling around. If it's lighted, it could take off."[25] Where do you see that dry kindling?

You see it all over the place. There's tremendous concern all over the country, all over the world, about repression, violence, domination, hierarchy, illegitimate authority. Take Bernie Sanders, for example. His proposed policies are policies that the public has supported for a long time, often with substantial majorities. In our dysfunctional system, public opinion couldn't be articulated in the political arena, but as soon as Sanders did so, he received substantial support. That's an indication of plenty of dry kindling.

10

ELECTIONS AND VOTING

CAMBRIDGE, MASSACHUSETTS (SEPTEMBER 9, 2016)

Let's talk about presidential elections. What's the earliest one you remember?

I remember 1936, when I was eight. The election was discussed a lot in school. In fact, I remember having big arguments with a classmate. He was in favor of Alf Landon and Frank Knox, the vice presidential candidate; his favorite slogan was "Landon Knox Out Roosevelt." In our circles, it was basically a hundred percent for FDR.

A couple of years ago, I was talking to an old friend, roughly my age, about family doctors. I couldn't remember our family doctor's name; the only name coming to my mind was "Roosevelt." After a while, I figured out why. Whenever my little brother had a cold, my mother assumed he was dying, and would call the doctor. Doctors used to come to the house in those days, and as soon as our doctor walked in, the whole mood would change.

He had a deep, mellifluous voice and an air of authority. It's all under control, everything's fine. My mother would immediately feel better.

Roosevelt used to give Fireside Chats on the radio, I think on Friday evenings. My mother was extremely nervous about everything. Of course, there was plenty to be nervous about—Hitler, the war. But as soon as Roosevelt started talking, with that quiet, serious voice of his, everything would calm down, very much like with the doctor. Which is why when I was thinking back to our doctor, Roosevelt came to mind. I still can't remember what his actual name was.

Do you recall a political period in which there was as much vitriol and recrimination, rancor, and rage as there is today?

The viciousness and vitriol of the personal attacks in some of this country's earliest elections are pretty shocking. But there's been nothing like what we have today. The current campaign is absolutely astonishing, partly because the major issues that humans face right now are simply not discussed. In the Republican primary, just about every single candidate denied anthropogenic climate warming—simply denied it, even though the facts are overwhelmingly clear. There was only one exception, John Kasich. He admitted that it's happening but said we shouldn't do anything about it, which is even worse.[1]

On nuclear war, too, there was virtually no discussion, just a few side comments here and there, usually framed in terms of Russian aggression and how we should respond to it. The fact that this is happening in the most powerful country in world history—by comparative standards, an educated, privileged country—is just incredible. I don't know how to even comment on the total denial of matters of such immense significance.

And the denial is not just abstract but is having real consequences. So, for example, the Paris climate negotiations, COP 21, though not as strong as they should be, were at least a step in the right direction, perhaps the basis for further action in the future. But the summit did not establish a treaty with verifiable goals—only an informal agreement. The reason was very clear: a treaty couldn't get through the Republican Congress. So here you have a political organization that is essentially saying, "Let's race to the precipice as fast as possible."

Donald Trump is being condemned for all sorts of things by liberal commentators, but not for the most important thing: his policies on the climate. He's calling for more fossil fuels, more coal plants, ending EPA regulations, possibly getting rid of the EPA altogether. He wants to dismantle the Paris agreements and stop giving support to developing countries for addressing climate change. That really is saying, "Let's race to the precipice." It's very serious. And the precipice is not far off. We're already close to—if not past—the limits that were proposed in the Paris discussions, a temperature increase of 1.5 degrees centigrade.

What is Trump saying that some people are receptive to? The standard explanation is that good-paying blue-collar jobs have disappeared, leaving a Rust Belt and angry workers in their wake.

There's surely something to that. The white working class has indeed been pretty much abandoned. The Democrats gave up

on white workers forty years ago. They don't offer them anything. The Republicans offer them even less. Or, rather, what
they, including Trump, offer them is a punch in the nose. The
health care system is scandalous? Let's make it worse. Trump's
proposed budget, which is basically Paul Ryan's, is devastating
for working people. There's increased spending on the military
and reduced taxes on the rich. There's essentially nothing left
for any moderately constructive part of the government.

But the Republicans do have a rhetorical style that makes it
sound as if they're working for working people. That's the rhetoric: "We're for you." It's not true. But what is true is that the
white working class feels that is everyone is against them. Nothing has been offered to them by either political party.

It's also true that close, careful analyses of Trump supporters
have shown a close correlation with authoritarian personalities—
patriarchal, authoritarian, racist, ultranationalist, and so on.[2]
That's significant as well.

But I wouldn't discount the idea that the white working class
is just extremely angry, and for good reason. They've been cast
to the winds. Real wages are roughly what they were in the
1960s—and have actually declined since the latest recession.[3]
Meanwhile, there's enormous wealth, which has become very
narrowly concentrated and very visible. Why shouldn't they be
angry? The solutions proposed by Trump will make the problems worse, but that's a different issue.

*I'm hesitant to use the word "fascist," because it's bandied about quite
promiscuously, but there's a whiff of fascism in the air. Is there any
credence to the notion that there are fascistic tendencies in Trump's
campaign?*

Something is coming out in the Trump campaign, but I think to call Trump a fascist is highly misleading. It attributes too much to him. I don't think he has any ideology, or pretty much anything except "Give me what I want, and somehow I'll do something for you. I'll make America great again." For example, he was asked in a debate what he would do about ISIS, and his answer was something like: Well, first I'm going to make America great again. And I'm going to call the generals, and they're going to give me a plan. And I have my plan, and I'll decide which plan is best, and then we'll go ahead and get rid of them. But first I'll make America great again.[4] Whatever that means, it's not fascism. It doesn't rise to that level. There's no plausible political category that it belongs to.

That slogan, "Make America great again," taps into a nostalgia for an imagined past, an America that never really existed.

Not entirely imagined. There was a time when the United States had a lot more clout on the international scene than it has today. As I've said, in the period right after the Second World War, the United States had overwhelming power. The United Nations was a tool in the hands of the U.S., a battering ram against Russia. That's all over.

Or take the International Monetary Fund, which is pretty much run by the U.S. Treasury. The IMF defers to the Europeans in dealing with Europe, but where the rest of the world is concerned, it also functioned as a tool of U.S. dominance. That, too, has changed radically since the Asian financial crisis in the late 1990s, when Asian countries have been refusing to accept IMF loans. And in the current millennium, Latin American countries have simply kicked out the IMF. They don't take IMF loans, with all their conditions. This is a radical shift, and another sign of the decline of U.S. power worldwide. It's not something that's in the headlines, but it's very significant. The power of the U.S. to dominate the world has declined.

On the other hand, if you look at U.S. corporations, their ownership of the world has remained very stable. Take your iPhone. Apple has a Taiwanese corporation, Foxconn, which runs huge assembly plants in China and employs Chinese workers in miserable conditions. The value added in China to the iPhone is very slight. Almost all the profit goes back to Apple and its subsidiaries. That means a large part of the gross domestic product of China is actually owned by Apple and other U.S. corporations. Even though U.S. power to run the world has declined, U.S. corporate power is still extraordinary.

And, of course, there's another respect in which there's some truth to the nostalgia. The 1950s and 1960s saw the highest growth rate in U.S. history. There was plenty of turmoil, conflict, and so on, but nevertheless there was a sense that the country was growing and developing. Young people, college students, could say, "I'm going to have a decent future." That's pretty much gone. There's a sense of hopelessness, of decline, a sense that we've reached our peak and the current generation won't have a better life than their parents. That kind of feeling is very widespread—and there is some reality to it. So the talk about "making America great again" is not totally empty.

Thomas Frank, who wrote What's the Matter with Kansas?—*in which he explained how working-class people actually vote against their own economic interests—told me recently that he's concerned that there's going to be a "smooth-talking" Trump the next time around, one who "will not piss people off."⁵ He regards that as an ominous prospect.*

We already have him. His name is Paul Ryan. I think he is more dangerous than Trump because he comes across as serious, thoughtful, with numbers and spreadsheets. But when you take a look at his programs, they're devastating. And, yes, he may run for president. On the other hand, four years of Trump could very well bring us to a tipping point on climate change, which would render other questions moot. That sounds apocalyptic, but if you take a look at the actual developments today and Trump's policies, assuming he implements them, it's a very dangerous mixture.

Hillary Clinton's comments at the American Legion National Convention in Cincinnati were saturated not just with boilerplate rhetoric but with constant use of the words "exceptional" and "exceptionalism." America is "the indispensable nation," she said.⁶ What do you think about Hillary Clinton?

Who doesn't use that terminology? It's absolutely standard in liberal Democratic rhetoric. There's simply no deviation from it. "The indispensable nation" phrase comes from Madeleine Albright and Bill Clinton.⁷ Look at what are called the liberal intellectual journals. Samantha Power, writing in the *New York Review of Books*, starts by favorably quoting Henry Kissinger, his usual rhetoric.⁸ But he's not quite correct, she says, because he doesn't recognize how important it is for us to enlist other countries in our marvelous efforts to do good all over the world. You find this everywhere.

Take, say, the simple word "aggression." Nobody has any hesitation in describing Vladimir Putin's takeover of Crimea or his actions in eastern Ukraine as aggression. That's aggression, no qualification. But have you ever heard the term used for the U.S. invasion of Iraq? Ever? Could any member of the articulate classes use that phrase in the mainstream? It's virtually inconceivable.

In fact, let's look again at Crimea. Whatever you think about it—the takeover was indeed an illegal act—does the United States have a better claim to Guantánamo Bay than Russia has to Crimea? No. We have much less of a claim. We took it at gunpoint over a century ago; it has no historical connections to the United States. Then we refused to give it back, even though Cuba, once it achieved independence from the United States, immediately demanded it back.

There's absolutely no justification for this. The United States is holding Guantánamo Bay only to impede and undermine Cuba's development. It's a major port. It's also the site of the worst human rights abuses in Cuba by far—in fact, the worst in the whole hemisphere, except maybe for Colombia. Is it aggression? Is that even a question?

The inability to face elementary facts is overwhelmingly true of the liberal, intellectual, articulate sectors of the population that are involved in commentary and discussion. And the rhetoric you quote from Clinton comes straight out of that.

In a recent Gallup poll, more than three-fourths of respondents said the United States is "on the wrong track."[9] Both Trump and Clinton are deeply unpopular. They're regarded as untrustworthy and dishonest.[10] Have you ever seen that before in a presidential campaign?

I have never seen anything quite like this, but it's an instance of something much broader. For several decades now, attitudes toward a number of key institutions have become very negative. Support for Congress has sometimes been down to literally single digits.[11] Banks are hated, corporations are hated. The Federal Reserve, which people don't know anything about, is hated. The government, of course, is hated. About the only institution that gets consistently pretty high ratings is the military, for other, separate reasons.

I think the distaste for the candidates is just a reflection of a pervasive malaise, a sense that everything is going wrong. And that's been increasing for quite a few years. There are polls going back decades showing that some 70 percent of the population believes that the government doesn't work for the people but, rather, is "run by a few big interests looking out for themselves."[12]

It's that feeling that Trump-style rhetoric about making America great again is attempting to appeal to. Everything's going wrong, everyone in the world is pushing us around, nobody is listening to us. Trump is now being violently denounced for saying that Putin is a stronger leader than Obama, because Putin gets what he wants and we don't.[13] But that's a feeling that people definitely have. Why isn't everyone in the Middle East doing what we want them to do? Why is China building bases in the South China Sea when we don't want them to?

Note, all of this assumes that the world is supposed to be ours: the South China Sea is supposed to be an American lake, the countries on the Russian border are ours to control. Of course, we would never accept, say, Russian forces in Mexico or Chinese aircraft carriers off the coast of California. The world is supposed to be ours, but people aren't listening. Why? What's happening? We've got to make America great again. All of this is part of the sense that we're going in the wrong direction.

At the same time, parents can see that their children do not have the kind of future that they themselves aspired to. In fact, social mobility in the United States, contrary to the Horatio Alger myth, is quite low compared to other developed societies.[14]

And, of course, there's tremendous poverty. Just travel around Boston. It looks like it's collapsing. I can remember the first time I went to Europe, in the early 1950s, when Europe was still recovering from the war. When you came back home, it was like returning to some kind of paradise as compared with Europe. Now it's the other way around. If you go to a poor country, like, say, Portugal, and come back here, it looks like you're returning to wreckage. The infrastructure is collapsing, the roads don't work, the bridges are falling, we don't have a health care system, schools are declining.

Back in the 1950s, John Kenneth Galbraith wrote about "private affluence and public squalor."[15] But now it's become much more dramatic. Private affluence for very narrow sectors has gone through the roof—and public squalor is almost everywhere you look. People can see that. They can see their children loaded down with school debt, lacking good opportunities. It's easy to blame foreigners, immigrants, people who are worse off than you are. That's Trump's line, his standard misdirection. But the background phenomena are real.

Student debt is at $1.3 trillion.[16] *I talked to a couple of young people in Boulder, Colorado. One is $40,000 in debt, the other one $100,000 in debt.*

Furthermore, student debt is designed so you can't get rid of it. You can't declare bankruptcy the way a business can, the way Trump has done over and over again, and then start over—the debt is with you forever. Even your Social Security can be garnished by the government to pay it. So it's a permanent burden—and a very strong disciplinary force. It means your options are limited.

In the 1960s, there was a general feeling of, Well, I can take off a couple of years and become an activist. Then I'll go back and pick up my life. That has changed. Now you're trapped. If you try to go back, you won't be able to. You don't have the same choices available to you. You've got to subordinate yourself to power.

What do you think of the Our Revolution organization formed in the wake of Bernie Sanders's defeat in the Democratic primaries?

Personally, I would prefer that he drop the word "revolution," because what he's proposing are mildly reformist initiatives. Which is not to say they're bad. We could use mildly reformist initiatives, but let's not give people the illusion that there's some dramatic change taking place. Sanders's proposals, which I favor, are basically a version of New Deal liberalism. They would not have surprised somebody like Dwight Eisenhower, who famously said that anyone who doesn't accept the New Deal doesn't belong in American politics.[17] That was a long time ago, and it's significant that in today's context Sanders is seen as so extreme. His proposals don't challenge or question the fundamental system of capitalist authoritarianism. That's not even under discussion, obviously. What he proposes is a good basis for doing better, but it's not a revolution.

There are now a number of popular movements developing out of the Sanders campaign, like the Brand New Congress group, which look quite sensible. They're also reformist, of course, but there's nothing wrong with that.

Can you talk about what's being called lesser-evil voting or strategic voting? How do you respond to people who say, "I want to vote my conscience"? Or, conversely, "Bring it on. I'm going to vote for Trump, because that will break the system and speed up the revolution"?

Lesser-evil voting should be simply called elementary rationality and elementary morality. If you live in a swing state, you have several choices. One choice is to vote for Trump because you honestly think he's better. Fine, nothing strategic there. Another possibility is to vote for Clinton because you think she's genuinely better or because you find Trump extremely dangerous. A third choice is to abstain or to vote for, say, Jill Stein, which works out to much the same thing. It's basic arithmetic: if you put one less vote in the Clinton column, you're making it easier for Trump to win. That's just a matter of numbers. If you think you're voting your conscience when you vote for Jill Stein, what you're saying is "My conscience prefers Trump."

It's the same with lesser-evil voting in general. The question about voting your conscience is, Do you care about what happens to the world or do you really only care about what you feel? If you only care about what you feel, you don't have any conscience, you're not a moral agent at all. So stop talking about conscience. If you care about the effect on others, then you'll ask, Well, what are the consequences of subtracting a vote from the only person who can plausibly beat Trump in this election? Again, it's simple arithmetic, elementary rationality.

Frankly, I think this entire discussion should take maybe five minutes of our time, period. If you think it through, the facts are obvious. After that, go on with the things that matter: activism, organizing, popular movements. That's what people should be engaged in. Maybe electoral work at lower levels, school boards, activism on the environment, and so on. That's what should be taking our time.

The fact that there's even discussion about this is an indication of how the Left is trapped by the propaganda system. We have an enormous propaganda system that tries to focus people's attention and energy on the quadrennial extravaganza. You shouldn't fall for that. The presidential campaign is not insignificant. But it's not the main story.

As to the idea of "bring it on"—let's heighten the contradictions—we've lived through that. The German Communist Party in 1932 was one example.

They voted for Hitler?

They said, Well, there is no difference between the Social Democrats and the Nazis, so let's just bring on the Nazis and we'll have a revolution. Yeah, there was a revolution, but not the one they were talking about.

The United States is not in a revolutionary situation by any means. And if you want it to get there, you're going to have to build the popular base for it. If you take the word "revolution" seriously, if you want to get rid of the capitalist system altogether, what you have to do is press the options within the system as far as you can. If the public agrees that it wants to go further, and the resistance of the system is too strong, then you have a revolution. But not before, not just from a small sect saying, "Let's go break windows in the banks."

What do you think about the Brexit vote and the rise of right-wing political parties in Europe?

First of all, it's not very clear that Brexit will be implemented. There are lots of possible arrangements that might be worked out. So I don't think it's a foregone conclusion that Britain will literally be out of the European Union. It may be in some respects, not in others.

There is a feeling in England, including on the British left, that Brexit will free them from the reactionary policies of the European Union, but that's misleading. For one thing, Britain has been a sponsor of those policies. So it's not that the European Union is imposing them on England. England has been supporting them and advancing them. Furthermore, what's been happening in England, from Margaret Thatcher to Tony Blair to David Cameron, is not an effect of the European Union, it's internal. So separating themselves from the European Union and the Brussels bureaucracy is not any kind of panacea for their problems. In fact, it could very well make them worse. It will weaken England and could leave it, even more than it is now, under U.S. influence.

The rise of right-wing parties is scary. Norbert Hofer, virtually a neo-Nazi figure, came very close to the presidency in Austria in 2016.[18] A large part of the reason is anti-immigrant feeling. In Denmark, you also see anti-immigrant, anti-Muslim feeling against the tiny fraction of the population who are not blond and blue-eyed.

Europe has always been much more racist than the United States, in my opinion. The racism hasn't been as visible because the populations have been fairly homogeneous. But as soon as the homogeneity begins to change, even slightly, the racism comes out into the open.

Take France, for example. The North African population there lives under awful conditions. They're a very small percentage of the total population, yet around 60 or 70 percent of the incarcerated in France are Muslim, mostly from North Africa.[19]

The rise of right-wing parties is largely a result of the willingness of the centrist parties, including the social democrats, to tolerate economic and social policies that are highly destructive. The austerity policies imposed by the "troika"—the European Commission, the IMF, and the European Central Bank—have been extremely detrimental. And there's good evidence that they were deliberately designed to undermine the welfare state.[20] As I've said, the purpose of austerity was not economic development—in fact, austerity is very harmful to that. The goal was to dismantle welfare state programs:

pensions, decent working conditions, regulations about labor rights, and so on.

And you see the right-wing backlash as a result?

Yes. But you have to trace that back to the willingness of the moderate and moderate-left parties to accept those policies.

11

CRISES AND ORGANIZING

Two major events occurred on November 8, 2016. One, obviously, was the U.S. election. Later that week, Der Spiegel *had a cover story headlined "The End of the World," with Trump depicted as a meteor hurtling toward Earth, his mouth wide open to devour it.[1] Can you talk about the second major event, which hardly got any coverage whatsoever?*

The second major event, which I believe was much more important than the U.S. elections, was a meeting of some two hundred countries in Marrakech, Morocco. The meeting, called COP 22—the twenty-second annual Conference of the Parties on climate change—was a follow-up to the Paris climate negotiations of December 2015, which ended up only making some verbal commitments on climate change, without fully spelling them out. The Marrakech conference was intended to define specific measures that countries would commit to, steps that might actually deal with this very urgent problem.

On November 7, the meeting opened in the normal way. The next day, November 8, the World Meteorological Organization presented its report on the current state of what's being called the Anthropocene, the geological epoch in which humans are drastically modifying the environment.[2] It was a pretty dire report. It found that Arctic ice had receded about 30 percent relative to its normal level, which means that there is less reflection and more absorption of solar rays, which amplifies global warming. It pointed out that COP 21 had set a goal of keeping the rise in global temperature below 1.5 degrees centigrade, but that we were already approaching that limit. A little more of an increase and we'd be over that cap. And a little higher than that would be irreversible. That was November 8.[3]

Then the meeting went into suspension while everyone watched the results of the U.S. election. On November 9, the conference basically collapsed. The only question on the minds of the delegates was, With the most important country in the world likely to pull out, will this project survive at all? The conference ended, and once again nothing much was achieved except some verbal commitments.[4]

The spectacle was pretty astounding. Here were nearly two hundred countries, practically the whole world, all hoping for a leader to point the way to decent survival. And who were they looking to? China. China is the leader that they're expecting will somehow save civilization from self-destruction. On the opposite side, there's one country they're afraid is going to destroy the whole thing—the supposed leader of the free world, the most powerful country in human history. It was quite a remarkable spectacle. And no less remarkable was the fact that there was no comment on it. What happened in Marrakech is pretty dramatic, and it may justify that headline in *Der Spiegel* about the end of the world, though not in the sense they intended.

Tens of millions of Bangladeshis are going to be climate refugees because of rising sea levels and extreme weather. Atiq Rahman, the top Bangladeshi climate scientist, says, "These migrants should have the right to move to the countries from which all these greenhouse gases are coming. Millions should be able to go to the United States."[5]

His comment was noted in a sentence buried in the *New York Times*. It should have been the headline, because it illustrates something quite significant about what's called the migrant crisis. Actually, Pope Francis put it pretty well. He said that migrants are not the cause of the crisis but the victims of the crisis.[6]

Why do we regard it as a crisis if, say, eight thousand miserable victims come to a rich, powerful country like Austria, with its eight million people? Other countries—much less wealthy ones—are accepting refugees. In Lebanon, perhaps 40 percent of the population is made up of refugees fleeing from one crime or another, recently from Iraq and Syria in particular.[7] Some of the refugees date from 1948 and the expulsion of Palestinians at the time of the establishment of Israel. That's Lebanon, a poor country which has not generated refugees itself.

Jordan, too, has accepted an immense number of refugees. And Syria was accepting huge numbers before its implosion. But rich countries, which have not only the capacity to absorb refugees but also a significant responsibility for creating the conditions from which they're fleeing, refuse to accept them. There's only one country in Europe that has a somewhat respectable record on this, Angela Merkel's Germany, which has accepted around eight hundred thousand.[8]

Aside from Germany, what both Europe and the U.S. are doing instead is trying to bribe someone else to keep the refugees. Europe tried to make a deal with Turkey to take care of the refugees fleeing from the Syrian conflagration, from Iraq, from Afghanistan—we're talking about a couple of million people. Erdoğan is not my favorite person, but when he refers to European hypocrisy, it's pretty hard to argue with him.[9]

The United States is doing exactly the same thing. When people flee from the northern tier of Central America, from the three countries that were devastated by Reaganite atrocities—El Salvador, Guatemala, and Honduras—we expect Mexico to stop them from reaching our borders. That's their job.

It's interesting that there is one country in the region which refugees are not fleeing: Nicaragua. Why? It's the one country that the U.S. government basically didn't ruin. The United States did carry out terror in Nicaragua, an attack on the government. But in the other three countries, terrorist forces supported by the United States *were* the government. I remember, in 1985 or so, I was in Managua with César Jérez, who was the rector of the Jesuit university, a leading church figure in Central America. He had had to flee from both Guatemala and El Salvador, and took refuge in Nicaragua. We were walking down the street one evening and stopped to talk with a policeman. Jérez pointed out to me that Nicaragua is the one country in the region where you don't have to be terrified when you see the police.

So, looking ahead to the effects of climate change: tens of millions of people are going to have to flee Bangladesh. Where are they going to go? And that's just the beginning of the story of climate refugees. The Himalayan glaciers are melting—that's the water supply for India and Pakistan. Already it's reported that there are about seventy-five million people in India who don't have access to clean drinking water.[10] What's going to happen when this number increases? One very likely scenario is conflict between India and Pakistan over diminishing water supplies on which both of them rely. These are nuclear weapons states. They're already virtually at war. Suppose a water war breaks out? It will turn into a nuclear war very quickly, and a major nuclear interchange might lead to what scientists have been warning about for decades: nuclear winter and global famine. In which case, it's basically over for all of us. So there's a point where the two major threats to survival begin to converge.

Once again, there's barely a word about it in the newspapers, just a scattered comment here and there. It's right in front of our eyes, yet all attention is given to Trump's tweets.

It's hard to find words to describe what's happening.

It certainly is. I can't.

In the wake of Trump's election, some have invoked Joe Hill's admonition: "Don't mourn, organize."

That's the right prescription. And we should recognize that there are real opportunities. Just take a look at the popular vote in the last election. Clinton won by about 2.7 million votes. More significant, among younger people, she won by a huge margin. And even more significant: in the primaries, among younger people, Sanders won by an even bigger margin. That's the hope for the future, if it can be organized and mobilized. And it can.

What would be some ways to organize against Trump?

First of all, I think some of Trump's initiatives should be supported—but supported in a way different from his proposals. One of his main agenda items is development of infrastructure in the United States, which is indeed in disastrous shape. So, yes, let's reconstruct the collapsing bridges, roads, water supplies, the energy system. That's a policy one should support.

However, we should bear in mind that this was actually Obama's policy, and it was killed by the Republican wrecking machine. When Obama came into office, the Republican leadership was quite open about their tactics: We'll make sure nothing happens, and maybe somehow we'll get back in power.[11] Now they're in power, so there is no reason why Congress would block this. Or, rather, there are some reasons, but the main one, to make the country ungovernable, is gone.

However, there are two major differences between Obama's original infrastructure proposal and Trump's. One is that Obama's was undertaken at a time when it would not have had a fiscal cost. Trump's is being undertaken in the context of fiscal policies adapted from Paul Ryan, who in my view is the most dangerous figure in U.S. politics. His policies involve a sharp tax cut for the rich and for the corporate sector, no new sources of revenue, and a huge buildup of what Trump calls our "depleted" military system, which, in fact, is already eons ahead of every other country in the world.[12] There's also some talk about middle-class relief, of course, but that's marginal at best.

The cost of those policies, according to various estimates, is quite high. So what's going to happen? It's predictable. Recall Dick Cheney's remark to Secretary of the Treasury Paul O'Neill in 2002, when the Bush spend-and-borrow programs were taking off. He said, "Ronald Reagan showed us that deficits don't matter."[13] He meant deficits that Republicans create to get popular support but that are then blamed on someone else—on Democrats, if possible. When Democrats are in power, deficits are a horror story, we can't allow them. But when Republicans are in power, if you take a look at the record—not Eisenhower, but everyone since—we can have huge deficits and leave it for somebody else to clean up the mess. That's one difference.

The second difference is that what Trump has described, at least so far, is not government stimulus but a taxpayer subsidy to private corporations to rebuild infrastructure that they would then own and profit from. And if that's carried out, we're not going to see reconstruction of the decaying water systems or installation of solar panels. What we're going to get is toll roads and the like. That's not the kind of infrastructure development the country needs, but it fits the dedication of the Republican Party, and Trump in particular, to private power and profits. The talk about the benefits for the rest of the population is just rhetoric. You use populist rhetoric to draw people in, and at the same time you kick them in the face.

Actually, financial elites are pretty happy right now, because Trump is doing exactly the opposite of his campaign pledges.

He is bringing the core of the financial establishment that he attacked—Goldman Sachs, JPMorgan Chase, and so on—into the White House. The stock market shows that the business community understands what's going on very well. Almost right away after the election, the stocks of energy corporations, banks, and military firms shot up. One of the most striking was Peabody Energy, the world's largest private coal producer, which was actually in bankruptcy proceedings at the time. Their stock value immediately went up by about 50 percent, and rose more later.[14]

The market surge is a reflection of what the business world expects Trump's actual policies to be. The spike for banks means financial deregulation, which lays the basis for the next financial crisis, when taxpayers will have to bail everyone out again. The spike for energy corporations means destruction of the environment. You know what good news for military corporations means. These are all portents for the future.

But all of this can be constrained, controlled, reversed, directed in other ways. Take, say, the white male working-class Trump voters. A fair number of them voted for Obama in 2008. They voted for Obama for a reason. The slogan of his campaign was "Hope and change," and these voters wanted change, quite rightly, quite seriously. In 2007, a year before the big financial crash, U.S. nonsupervisory workers had lower wages than in 1979.[15] So the call for change made a lot of sense. Hope, too, of course.

Well, they didn't get any hope, and they didn't get any change. So now they voted for a con man who made the same promises. But that's an opportunity to provide a real and meaningful program for actual hope and change. That could bring together many of the Trump voters and the Sanders voters, for example. These are opportunities that can be pursued.

Let's talk about what's called the "gig economy," the vanishing of steady nine-to-five jobs that come with some benefits, maybe sick leave, maybe a couple of weeks' paid vacation. Such jobs are becoming distant memories as this economy of temporary work and on-demand hiring develops. Workers are supposed to feel good that they're now independent contractors. I guess you see a lot of this in universities and colleges, where there are now so many adjuncts and teaching assistants.

One of the things that happened during the neoliberal era is the increased imposition of a business model on universities, which means policies of the kind you describe: hiring temporary workers with no security, rotten working conditions, and very low wages. A tenured faculty member has job protections, but you can fire adjuncts and assistants anytime you want. It's a loss for working people and it's a loss for students, but it's good for the bottom line.

These workers are being called "independent contractors," although they're not that, of course. But the effect is to separate workers from one another, to focus them on their individual situations, which are precarious. They have no security, they can't plan, which means they're not likely to organize or to act as participants in a functioning democratic society. This is all good for concentrated power. Elites don't want a functioning democratic society. They want a society in which people are frightened, intimidated, inactive, fearing for the next paycheck.

What about the related phenomenon of private car services such as Uber and Lyft? Some people find them very convenient, but they're a real threat to the livelihood of taxi drivers.

They certainly are. The taxi drivers are people who have worked hard; they've often paid a lot for a taxi medallion. That's their life's investment. And it's being undermined and taken away by part-timers. You can understand the part-timers—they need the work. But you can also easily understand the concern of working people who organized, perhaps unionized, and had developed a life that offers them some degree of security and benefits.

During the presidential campaign, there was a lot of focus on the middle class but virtually no mention of the poor, other than by Bernie Sanders. With the proposed cuts to Medicaid, Head Start, Pell Grants, food stamps, and a number of other anti-poverty programs, the poor are going to get the short end of the stick once again. Why are they so marginalized politically?

Because they're powerless, unorganized, without resources. Immigrants, single mothers—they're just trying to survive. And a lot of it is racialized. For these people to organize and act is extremely difficult, so it's easy to disregard them.

There's a very interesting study by Arlie Hochschild, a sociologist who lived for five years in a Tea Party stronghold in Louisiana, trying to understand the point of view of the people there.[16] She wanted to solve what she describes as a great paradox, the same one Thomas Frank and others have talked about: Why are people voting for politicians who are out to destroy them? These people live around the bayou, they hunt and fish, and have memories of a wonderful life in a lush environment. Now it's all being destroyed by petrochemical plants. There is a "cancer alley" there now. The people are angry. And yet they vote for the candidate who plans to increase the damage.

Hochschild offers an interesting image. She says the people she is writing about see themselves as standing in line. They've worked hard all their lives, they did all the proper things: they went to church, they studied the Bible, they had families, the husband went to work, the wife stayed home and took care of the kids. They did everything right, just as their parents did, and they were slowly moving forward in the line. That's the American Dream. Then, all of a sudden, the line stalled, and now they're about where they were twenty-five or thirty years ago, or even further back.

Ahead of them, some people are going off into the stratosphere. But that doesn't bother them, because the essence of the American Dream is that if you have merit, you're supposed to benefit. What bothers them is the people behind them, the poor, the "worthless" ones who haven't worked hard or done anything right. And all of a sudden, the federal government comes along and takes those people and puts them ahead in the line.

It's a myth, but there are enough particles of truth scattered in the myth that it can be made to look credible. Remember the "welfare queens," who were supposedly going to the welfare office in their limousines and stealing your pay? Now the immigrants are being favored, along with the African Americans, the Latinos, the assorted scroungers. It's "the makers" versus "the takers." And the federal government is to blame for putting the scroungers ahead of the hardworking, decent, sober folk.

So they hate the federal government. They're being devastated by pollution from the petrochemical plants—but their image of the EPA is some guy in a business suit who comes down there and tells them they're not allowed to fish but doesn't do anything about those plants. And all this fits together with religious beliefs, elements of white supremacy, patriarchal commitments. There's an internal logic that makes some sense, even though, in the broader context, it's self-destructive.

But here, too, change is possible. And there are opportunities for organizing.

I was looking recently at the Populist Party platform of 1892 in Omaha. It called for nationalization of railroads and steamship lines, nationalization of telephone and telegraph systems.[17] How would you characterize what is called populism today?

That was the original populist movement. It was a movement of radical farmers that started in Texas and spread to Kansas and throughout the Midwest. In that period, agrarian society was being marginalized in favor of state capitalist industrialization. Farmers were being strangled by northern finance, by bankers, by merchants, and forced to work within a highly exploitative system. These people maintained the individualist republican conception that a person is free only if no one is giving them orders. So wage labor was considered not all that different from slavery. It was temporary, you could get out of it, but it was a form of bondage.

That background gave rise to a radical populist movement, which made efforts to unify with the growing industrial organized working class. The populists were crushed, partially by force. But theirs was the most democratic movement in U.S. history. It later degenerated into racism, xenophobia, and anti-Semitism. But its origins were highly progressive, and it helped shape the progressive movement of the early twentieth century.

What's called populism today is something quite different. It means opposition to established institutions, which can take all kinds of forms. It could be could be highly leftist and progressive, or it could be neo-Nazism.

Were you impressed by the resistance that the Standing Rock Sioux mobilized against the Keystone XL oil pipeline?

Yes, though the Trump administration has already said it will approve the existing plans for the pipeline route.[18]

Still, a broad alliance coalesced around one of the most oppressed minorities in this country, the Native Americans.

Yes, many issues converged: Native rights, treaties that are again being broken, local environmental effects, water rights. And then there's the broader point: the oil should stay in the ground. Unfortunately, even if the pipeline is rerouted, which doesn't look likely, the oil is not going to stay in the ground.

12

THE TRUMP PRESIDENCY

E-MAIL EXCHANGE (JUNE 20, 2017)

You have spoken about the difference between Trump's buffoonery, which gets endlessly covered by the media, and the actual policies he is striving to enact, which receive less attention. Do you think he has any coherent economic, political, or international policy goals? What has Trump actually managed to accomplish in his first months in office?

There is a diversionary process under way, perhaps by design, perhaps just a natural result of the propensities of the figure at center stage and those doing the work behind the curtains.

At one level, Trump's antics ensure that attention is focused on him, and it makes little difference how. Who even remembers the charge that millions of illegal immigrants voted for Clinton, depriving the pathetic little man of his Grand Victory?[1] Or the accusation that Obama had wiretapped Trump Tower?[2] The claims themselves don't really matter. It's enough that attention is diverted from what is happening in the background. There, out of the spotlight, the most savage fringe of the Republican Party is carefully advancing policies designed to enrich their true constituency: the Constituency of private power and wealth, "the masters of mankind," to borrow Adam Smith's phrase.

These policies will harm the irrelevant general population and devastate future generations, but that's of little concern to the Republicans. They've been trying to push through similarly destructive legislation for years. Paul Ryan, for example, has long been advertising his ideal of virtually eliminating the federal government, apart from service to the Constituency—though in the past he's wrapped his proposals in spreadsheets so they would look wonkish to commentators. Now, while attention is focused on Trump's latest mad doings, the Ryan gang and the executive branch are ramming through legislation and orders that undermine workers' rights, cripple consumer protections, and severely harm rural communities. They seek to devastate health programs, revoking the taxes that pay for them in order to further enrich their Constituency, and to eviscerate the Dodd-Frank Act, which imposed some much-needed constraints on the predatory financial system that grew during the neoliberal period.[3]

That's just a sample of how the wrecking ball is being wielded by the newly empowered Republican Party. Indeed, it is no longer a political party in the traditional sense. Conservative political analysts Thomas Mann and Norman Ornstein have described it more accurately as a "radical insurgency," one that has abandoned normal parliamentary politics.[4]

Much of this is being carried out stealthily, in closed sessions, with as little public notice as possible. Other Republican policies are more open, such as pulling out of the Paris climate agreement, thereby isolating the U.S. as a pariah state that refuses to participate in international efforts to confront looming environmental disaster. Even worse, they are intent on maximizing the use of fossil fuels, including the most dangerous; dismantling regulations; and sharply cutting back on research and development of alternative energy sources, which will soon be necessary for decent survival.

The reasons behind the policies are a mix. Some are simply service to the Constituency. Others are of little concern to the "masters of mankind" but are designed to hold on to segments of the voting bloc that the Republicans have cobbled together, since Republican policies have shifted so far to the right that their actual proposals would not attract voters. For example, terminating support for family planning is not service to the Constituency. Indeed, that group may mostly support family planning. But terminating that support appeals to the evangelical Christian base—voters who close their eyes to the fact that they are effectively advocating more unwanted pregnancies and, therefore, increasing the frequency of resort to abortion, under harmful and even lethal conditions.

Not all of the damage can be blamed on the con man who is nominally in charge, on his outlandish appointments, or on the congressional forces he has unleashed. Some of the most dangerous developments under Trump trace back to Obama initiatives—initiatives passed, to be sure, under pressure from the Republican Congress.

The most dangerous of these has barely been reported. A very important study in the *Bulletin of the Atomic Scientists*, published in March 2017, reveals that the Obama nuclear weapons modernization program has increased "the overall killing power of existing US ballistic missile forces by a factor of roughly three—and it creates exactly what one would expect to see, if a nuclear-armed state were planning to have the capacity to fight and win a nuclear war by disarming enemies with a surprise first strike."[5] As the analysts point out, this new capacity undermines the strategic stability on which human survival depends. And the chilling record of near disaster and reckless behavior of leaders in past years only shows how fragile our survival is. Now this program is being carried forward under Trump. These developments, along with the threat of environmental disaster, cast a dark shadow over everything else—and are barely discussed, while attention is claimed by the performances of the showman at center stage.

Whether Trump has any idea what he and his henchmen are up to is not clear. Perhaps he is completely authentic: an ignorant, thin-skinned megalomaniac whose only ideology is himself. But what is happening under the rule of the extremist wing of the Republican organization is all too plain.

Do you see any encouraging activity on the Democrats' side? Or is it time to begin thinking about a third party?

There is a lot to think about. The most remarkable feature of the 2016 election was the Bernie Sanders campaign, which broke the pattern set by over a century of U.S. political history. A substantial body of political science research convincingly establishes that elections are pretty much bought; campaign funding alone is a remarkably good predictor of electability, for Congress as well as for the presidency. It also predicts the decisions of elected officials. Correspondingly, a considerable majority of the electorate—those lower on the income scale—are effectively disenfranchised, in that their representatives disregard their preferences. In this light, there is little surprise in the victory of a billionaire TV star with substantial media backing: direct backing from the leading cable channel, Rupert Murdoch's Fox, and from highly influential right-wing talk radio; indirect but lavish backing from the rest of the major media, which was entranced by Trump's antics and the advertising revenue that poured in.

The Sanders campaign, on the other hand, broke sharply from the prevailing model. Sanders was barely known. He had virtually no support from the main funding sources, was ignored or derided by the media, and labeled himself with the scare word "socialist." Yet he is now the most popular political figure in the country by a large margin.[6]

At the very least, the success of the Sanders campaign shows that many options can be pursued even within the stultifying two-party framework, with all of the institutional barriers to breaking free of it. During the Obama years, the Democratic Party disintegrated at the local and state levels. The party had largely abandoned the working class years earlier, even more so with Clinton trade and fiscal policies that undermined U.S. manufacturing and the fairly stable employment it provided.

There is no dearth of progressive policy proposals. The

program developed by Robert Pollin in his book *Greening the Global Economy* is one very promising approach.[7] Gar Alperovitz's work on building an authentic democracy based on worker self-management is another.[8] Practical implementations of these approaches and related ideas are taking shape in many different ways. Popular organizations, some of them outgrowths of the Sanders campaign, are actively engaged in taking advantage of the many opportunities that are available.

At the same time, the established two-party framework, though venerable, is by no means graven in stone. It's no secret that in recent years, traditional political institutions have been declining in the industrial democracies, under the impact of what is called "populism." That term is used rather loosely to refer to the wave of discontent, anger, and contempt for institutions that has accompanied the neoliberal assault of the past generation, which led to stagnation for the majority alongside a spectacular concentration of wealth in the hands of a few.

Functioning democracy erodes as a natural effect of the concentration of economic power, which translates at once to political power by familiar means, but also for deeper and more principled reasons. The doctrinal pretense is that the transfer of decision-making from the public sector to the "market" contributes to individual freedom, but the reality is different. The transfer is from public institutions, in which voters have some say, insofar as democracy is functioning, to private tyrannies— the corporations that dominate the economy—in which voters have no say at all. In Europe, there is an even more direct method of undermining the threat of democracy: placing crucial decisions in the hands of the unelected troika—the

International Monetary Fund, the European Central Bank, and the European Commission—which heeds the northern banks and the creditor community, not the voting population.

These policies are dedicated to making sure that society no longer exists, Margaret Thatcher's famous description of the world she perceived—or, more accurately, hoped to create: one where there is no society, only individuals. This was Thatcher's unwitting paraphrase of Marx's bitter condemnation of repression in France, which left society as a "sack of potatoes," an amorphous mass that cannot function.[9] In the contemporary case, the tyrant is not an autocratic ruler—in the West, at least—but concentrations of private power.

The collapse of centrist governing institutions has been evident in elections: in France in mid-2017 and in the United States a few months earlier, where the two candidates who mobilized popular forces were Sanders and Trump—though Trump wasted no time in demonstrating the fraudulence of his "populism" by quickly ensuring that the harshest elements of the old establishment would be firmly ensconced in power in the luxuriating "swamp."

These processes might lead to a breakdown of the rigid American system of one-party business rule with two competing factions, with varying voting blocs over time. They might provide an opportunity for a genuine "people's party" to emerge, a party where the voting bloc is the actual constituency, and the guiding values merit respect.

Trump's first foreign trip was to Saudi Arabia. What significance do you see in that, and what does it mean for broader Middle East policies? And what do you make of Trump's animus toward Iran?

Saudi Arabia is the kind of place where Trump feels right at home: a brutal dictatorship, miserably repressive (notoriously so for women's rights, but in many other areas as well), the leading producer of oil (now being overtaken by the United States), and with plenty of money. The trip produced promises of massive weapons sales—greatly cheering the Constituency—and vague intimations of other Saudi gifts. One of the consequences was that Trump's Saudi friends were given a green light to escalate their disgraceful atrocities in Yemen and to discipline Qatar, which has been a shade too independent of the Saudi masters. Iran is a factor there. Qatar shares a natural gas field with Iran and has commercial and cultural relations with it, frowned upon by the Saudis and their deeply reactionary associates.

Iran has long been regarded by U.S. leaders, and by U.S. media commentary, as extraordinarily dangerous, perhaps the most dangerous country on the planet. This goes back to well before Trump. In the doctrinal system, Iran is a dual menace: it is the leading supporter of terrorism, and its nuclear programs pose an existential threat to Israel, if not the whole world. It is so dangerous that Obama had to install an advanced air defense system near the Russian border to protect Europe from Iranian nuclear weapons—which don't exist, and which, in any case, Iranian leaders would use only if possessed by a desire to be instantly incinerated in return.

That's the doctrinal system. In the real world, Iranian support for terrorism translates to support for Hezbollah, whose major crime is that it is the sole deterrent to yet another destructive Israeli invasion of Lebanon, and for Hamas, which won a free election in the Gaza Strip—a crime that instantly elicited harsh sanctions and led the U.S. government to prepare a military coup. Both organizations, it is true, can be charged with terrorist acts, though not anywhere near the amount of terrorism that stems from Saudi Arabia's involvement in the formation and actions of jihadi networks.

As for Iran's nuclear weapons programs, U.S. intelligence has confirmed what anyone can easily figure out for themselves: if they exist, they are part of Iran's deterrent strategy. There is also the unmentionable fact that any concern about Iranian weapons of mass destruction (WMDs) could be alleviated by the simple means of heeding Iran's call to establish a WMD-free zone in the Middle East. Such a zone is strongly supported by the Arab states and most of the rest of the world and is blocked primarily by the United States, which wishes to protect Israel's WMD capabilities.

Since the doctrinal system falls apart on inspection, we are left with the task of finding the true reasons for U.S. animus toward Iran. Possibilities readily come to mind. The United States and Israel cannot tolerate an independent force in a region that they take to be theirs by right. An Iran with a nuclear deterrent is unacceptable to rogue states that want to rampage however they wish throughout the Middle East. But there is more to it than that. Iran cannot be forgiven for overthrowing the dictator installed by Washington in a military coup in 1953, a coup that destroyed Iran's parliamentary regime and its unconscionable belief that Iran might have some claim on its own natural resources. The world is too complex for any simple description, but this seems to me the core of the tale.

It also wouldn't hurt to recall that in the past six decades, scarcely a day has passed when Washington was not tormenting

Iranians. After the 1953 military coup came U.S. support for a dictator described by Amnesty International as a leading violator of fundamental human rights. Immediately after his overthrow came the U.S.-backed invasion of Iran by Saddam Hussein, no small matter. Hundreds of thousands of Iranians were killed, many by chemical weapons. Reagan's support for his friend Saddam was so extreme that when Iraq attacked a U.S. ship, the USS *Stark*, killing thirty-seven American sailors, it received only a light tap on the wrist in response. Reagan also sought to blame Iran for Saddam's horrendous chemical warfare attacks on Iraqi Kurds.

Eventually, the United States intervened directly in the Iran-Iraq War, leading to Iran's bitter capitulation. Afterward, George H. W. Bush invited Iraqi nuclear engineers to the United States for advanced training in nuclear weapons production—an extraordinary threat to Iran, quite apart from its other implications.[10] And, of course, Washington has been the driving force behind harsh sanctions against Iran that continue to the present day.

Trump, for his part, has joined the harshest and most repressive dictators in shouting imprecations at Iran. As it happens, Iran held an election during his Middle East travel extravaganza—an election which, however flawed, would be unthinkable in the land of his Saudi hosts, who also happen to be the source of the radical Islamism that is poisoning the region. But U.S. animus against Iran goes far beyond Trump himself. It includes those regarded as the "adults" in the Trump administration, like James "Mad Dog" Mattis, the secretary of defense. And it stretches a long way into the past.

What are the strategic issues where Korea is concerned? Can anything be done to defuse the growing conflict?

Korea has been a festering problem since the end of World War II, when the hopes of Koreans for unification of the peninsula were blocked by the intervention of the great powers, the United States bearing primary responsibility.

The North Korean dictatorship may well win the prize for brutality and repression, but it is seeking and to some extent carrying out economic development, despite the overwhelming burden of a huge military system. That system includes, of course, a growing arsenal of nuclear weapons and missiles, which pose a threat to the region and, in the longer term, to countries beyond—but its function is to be a deterrent, one that the North Korean regime is unlikely to abandon as long as it remains under threat of destruction.

Today, we are instructed that the great challenge faced by the world is how to compel North Korea to freeze these nuclear and missile programs. Perhaps we should resort to more sanctions, cyberwar, intimidation; to the deployment of the Terminal High Altitude Area Defense (THAAD) anti-missile system, which China regards as a serious threat to its own interests; perhaps even to direct attack on North Korea—which, it is understood, would elicit retaliation by massed artillery, devastating Seoul and much of South Korea even without the use of nuclear weapons.

But there is another option, one that seems to be ignored: we could simply accept North Korea's offer to do what we are demanding. China and North Korea have already proposed that North Korea freeze its nuclear and missile programs. The proposal, though, was rejected at once by Washington, just as it had been two years earlier, because it includes a quid pro quo: it calls on the United States to halt its threatening military exercises on North Korea's borders, including simulated nuclear-bombing attacks by B-52s.[11]

The Chinese–North Korean proposal is hardly unreasonable. North Koreans remember well that their country was literally flattened by U.S. bombing, and many may recall how U.S. forces bombed major dams when there were no other targets left.[12] There were gleeful reports in American military publications about the exciting spectacle of a huge flood of water wiping out the rice crops on which "the Asian" depends for survival.[13] They are very much worth reading, a useful part of historical memory.

The offer to freeze North Korea's nuclear and missile programs in return for an end to highly provocative actions on North Korea's border could be the basis for more far-reaching negotiations, which could radically reduce the nuclear threat and perhaps even bring the North Korea crisis to an end. Contrary to much inflamed commentary, there are good reasons to think such negotiations might succeed. Yet even though the North Korean programs are constantly described as perhaps the greatest threat we face, the Chinese–North Korean proposal is unacceptable to Washington, and is rejected by U.S. commentators with impressive unanimity. This is another entry in the shameful and depressing record of near-reflexive preference for force when peaceful options may well be available.

The 2017 South Korean elections may offer a ray of hope.

Newly elected President Moon Jae-in seems intent on reversing the harsh confrontationist policies of his predecessor.[14] He has called for exploring diplomatic options and taking steps toward reconciliation, which is surely an improvement over the angry fist-waving that might lead to real disaster.

You have in the past expressed concern about the European Union. What do you think will happen as Europe becomes less tied to the U.S. and the U.K.?

The E.U. has fundamental problems, notably the single currency with no political union. It also has many positive features. There are some sensible ideas aimed at saving what is good and improving what is harmful. Yanis Varoufakis's DiEM25 initiative for a democratic Europe is a promising approach.[15]

The U.K. has often been a U.S. surrogate in European politics. Brexit might encourage Europe to take a more independent role in world affairs, a course that might be accelerated by Trump policies that increasingly isolate us from the world. While he is shouting loudly and waving an enormous stick, China could take the lead on global energy policies while extending its influence to the west and, ultimately, to Europe, based on the Shanghai Cooperation Organization and the New Silk Road.

That Europe might become an independent "third force" has been a matter of concern to U.S. planners since World War II. There have long been discussions of something like a Gaullist conception of Europe from the Atlantic to the Urals or, in more recent years, Gorbachev's vision of a common Europe from Brussels to Vladivostok.

Whatever happens, Germany is sure to retain a dominant role in European affairs. It is rather startling to hear a conservative German chancellor, Angela Merkel, lecturing her U.S. counterpart on human rights, and taking the lead, at least for a time, in confronting the refugee issue, Europe's deep moral crisis. On the other hand, Germany's insistence on austerity and paranoia about inflation and its policy of promoting exports by limiting domestic consumption have no slight responsibility for Europe's economic distress, particularly the dire situation of the peripheral economies. In the best case, however, which is not beyond imagination, Germany could influence Europe to become a generally positive force in world affairs.

What do you make of the conflict between the Trump administration and the U.S. intelligence communities? Do you believe in the "deep state"?

There is a national security bureaucracy that has persisted since World War II. And national security analysts, in and out of government, have been appalled by many of Trump's wild forays. Their concerns are shared by the highly credible experts who set the Doomsday Clock, advanced to two and a half minutes to midnight as soon as Trump took office—the closest it has been to terminal disaster since 1953, when the U.S. and USSR exploded thermonuclear weapons.[16] But I see little sign that it goes beyond that, that there is any secret "deep state" conspiracy.

To conclude, as we look forward to your eighty-ninth birthday, I wonder: Do you have a theory of longevity?

Yes, it's simple, really. If you're riding a bicycle and you don't want to fall off, you have to keep going—fast.

NOTES

1. STATE SPYING AND DEMOCRACY

1. James Risen and Nick Wingfield, "Silicon Valley and Spy Agency Bound by Strengthening Web," *New York Times*, 20 June 2013, A1.

2. Alfred W. McCoy, *Policing America's Empire: The United States, the Philippines, and the Rise of the Surveillance State* (Madison: University of Wisconsin Press, 2009).

3. Glenn Greenwald and Ewen MacAskill, "NSA Prism Program Taps In to User Data of Apple, Google and Others," *Guardian*, 7 June 2013. Available online at: https://www.theguardian.com/world/2013/jun/06/us-tech-giants-nsa-data.

4. Bruce Schneier, "Online Nationalism," *MIT Technology Review* 116:3 (May–June 2013): 12.

5. Noam Chomsky et al., *Trials of the Resistance* (New York: New York Review, 1970).

6. Charlie Savage, "Drone Strikes Turn Allies into Enemies, Yemeni Says," *New York Times*, 23 April 2013. The testimony of Farea al-Muslimi is available online at: "Yemeni Man Brings the Horror of Drone Strikes Home to the US Senate," *Independent* (London), 24 April 2013.

7. Adam Liptak, "Justices Uphold a Ban on Aiding Terror Groups," *New York Times*, 22 June 2010, A1. The published opinion for *Attorney General*

Holder v. Humanitarian Law Project is available online at: https://www
.supremecourt.gov/opinions/09pdf/08-1498.pdf.

8. Caitlin Dewey, "Why Nelson Mandela Was on a Terrorism Watch List in 2008," *Washington Post*, 7 December 2013. Available online at: https:// www.washingtonpost.com/news/the-fix/wp/2013/12/07/why-nelson -mandela-was-on-a-terrorism-watch-list-in-2008/.

9. Joyce Battle, ed., "Shaking Hands with Saddam Hussein: The U.S. Tilts Toward Iraq, 1980–1984," National Security Archive Electronic Briefing Book no. 82, February 25, 2003. Available online at: http://nsarchive.gwu .edu/NSAEBB/NSAEBB82/.

10. See Gallup poll data available at: http://www.gallup.com/poll/1714/taxes .aspx.

11. Newsweek poll/Princeton Survey Research Associates International, "Obama/Muslims: Final Topline Results," 27 August 2010. Available online at: http://nw-assets.s3.amazonaws.com/pdf/1004-ftop.pdf.

12. Ezra Klein, "Rand Paul: Obama Is Working with 'Anti-American Globalists Plot[ting] Against Our Constitution,' " *Washington Post*, 11 May 2013. Available online at: https://www.washingtonpost.com/news/wonk/wp /2013/05/11/rand-paul-obama-is-working-with-anti-american-globalists -plotting-against-our-constitution/.

13. William Jay, *The Life of John Jay* (New York: J. & J. Harper, 1833), vol. 1, ch. 3.

14. James Madison, quoted in *Notes of the Secret Debates of the Federal Convention of 1787, Taken by the Late Hon Robert Yates, Chief Justice of the State of New York, and One of the Delegates from That State to the Said Convention*, 26 June 1787. Available online at: http://avalon.law.yale.edu/18th_century /yates.asp.

2. A TOUR OF THE MIDDLE EAST

1. Noam Chomsky, "Cambodia," special supplement, *New York Review of Books* 14, no. 11 (June 4, 1970): 39–50.

2. Robert Fisk, "Iran to Send 4,000 Troops to Aid President Assad Forces in Syria," *Independent* (London), 16 June 2013.

3. Barak Ravid and Reuters, "WikiLeaks: Israel Weapons Manufacturer Listed as Site Vital to U.S. Interests," *Haaretz*, 6 December 2010. Available online at: http://www.haaretz.com/israel-news/wikileaks-israel-weapons -manufacturer-listed-as-site-vital-to-u-s-interests-1.329222.

4. For analysis, see Noam Chomsky, *The Culture of Terrorism*, 3rd ed. (Chicago: Haymarket Books, 2015).

5. Elise Ackerman, "Israeli Software Maker Varonis Systems Files for IPO," *Forbes*, 23 October 2013.

6. Emily Greenhouse, "The Armenian Past of Taksim Square," *New Yorker*, 28 June 2013. Available online at: http://www.newyorker.com/culture/culture-desk/the-armenian-past-of-taksim-square.

7. Vincent Boland, "Journalist Killed by Gunman in Istanbul," *Financial Times*, 19 January 2007. Available online at: https://www.ft.com/content/6b6f26ea-a7d0-11db-b448-0000779e2340.

8. Sebnem Arsu and Ceylan Yeginsu, "Turkish Leader Offers Referendum on Park at Center of Protests," *New York Times*, 12 June 2013. Available online at: http://www.nytimes.com/2013/06/13/world/europe/taksim-square-protests-istanbul-turkey.html.

9. Amnesty International, "Turkey Accused of Gross Human Rights Violations in Gezi Park Protests," 2 October 2013. Available online at: https://www.amnesty.org/en/latest/news/2013/10/turkey-accused-gross-human-rights-violations-gezi-park-protests/.

10. Committee to Protect Journalists, "Turkey's Crackdown Propels Number of Journalists in Jail Worldwide to Record High," 13 December 2013. Available online at: https://www.cpj.org/reports/2016/12/journalists-jailed-record-high-turkey-crackdown.php.

11. David Hume, "Of National Character," I.XXI.14. Available online at: http://www.econlib.org/library/LFBooks/Hume/hmMPL21.html.

3. POWER SYSTEMS DO NOT GIVE GIFTS

1. John Bellamy Foster and Robert W. McChesney, *The Endless Crisis: How Monopoly Finance-Capital Produces Stagnation and Upheaval from the U.S.A. to China* (New York: Monthly Review Press, 2012).

2. See *Manufacturing Consent*, directed by Mark Achbar and Peter Wintonick (Zeitgeist Films, 1993), and the accompanying book of the same title, published by Black Rose Books in Montreal in 1994.

3. See Rudolf Rocker, *Anarcho-Syndicalism: Theory and Practice* (1938; repr., Oakland: AK Press, 2004), with a preface by Noam Chomsky.

4. Ibid., 74.

5. Noam Chomsky, "Notes on Anarchism," in *The Essential Chomsky*, ed. Anthony Arnove (New York: New Press, 2008), 104.

6. Chomsky, "Language and Freedom," in *The Essential Chomsky*, 89.

7. Steve Horn, "ALEC Model Bill Behind Push to Require Climate Denial Instruction in Schools," Desmog Blog, 26 January 2012. Available online at: https://www.desmogblog.com/alec-model-bill-behind-push-require-climate-denial-instruction-schools.

8. Lisa Graves, "ALEC Exposed: The Koch Connection," *Nation*, 12 July 2011. Available online at https://www.thenation.com/article/alec-exposed-koch-connection/.

9. See Adam Davidson, "How AIG Fell Apart," Reuters, 18 September 2008. Available online at: http://www.reuters.com/article/us-how-aig-fell-apart-idUSMAR85972720080918.

10. Ellen Cantarow, "No Pipe Dream: Is Fracking About to Arrive on Your Doorstep?" TomDispatch.com, 30 January 2014. Available online at: http://www.tomdispatch.com/blog/175800/tomgram%3A_ellen_cantarow,_the_frontlines_of_fracking.

11. IPCC, *Climate Change*, Fifth Assessment Report, p. 15. Percentage figures of certainty are given on p. 4, footnote 2. See a comparison to the 2007 report by Dana Nuccitelli, "Global Warming: Why Is IPCC Report So Certain About the Influence of Humans?" *Guardian*, 27 September 2013. Available online at: https://www.theguardian.com/environment/climate-consensus-97-per-cent/2013/sep/27/global-warming-ipcc-report-humans.

12. Glenn Scherer, "Climate Science Predictions Prove Too Conservative," *Scientific American*, 6 December 2012. Available online at: http://www.scientificamerican.com/article/climate-science-predictions-prove-too-conservative/.

13. Andres Schipani, "Ecuador Admits Defeat in Plan to Keep Oil in the Ground for a Fee," *Financial Times*, 16 August 2013. Available online at: https://www.ft.com/content/99e438ae-0691-11e3-ba04-00144feab7de.

14. Amos 7:14.

15. Telegram from the Department of State to the Embassy in South Africa, 25 October 1961, Department of State, Central Files, 611.70X/10–2461; Memorandum of Conversation Between South African Representatives and the Security Council, 17 July 1963, Department of State, Central Files, POL 1 S AFR. See entire declassified record from this period, available online at the U.S. Department of State Archive, https://2001-2009.state.gov/r/pa/ho/frus/kennedyjf/50766.htm.

16. "Hill Overrides Veto of South Africa Sanctions," 1986, *CQ Almanac*, https://library.cqpress.com/cqalmanac/document.php?id=cqal86-1149011.

17. Caitlin Dewey, "Why Nelson Mandela Was on the Terrorism Watch List in 2008," *Washington Post*, 7 December 2013.

18. Richard Boudreaux, "Mandela Lauds Castro as Visit to Cuba Ends," *Los Angeles Times*, 28 July 1991. Available online: http://articles.latimes.com/1991-07-28/news/mn-519_1_leader-nelson-mandela.

19. William M. LeoGrande and Peter Kornbluh, *Back Channel to Cuba: The Hidden History of Negotiations Between Washington and Havana* (University of North Carolina Press: 2015), 145–48.

20. Noam Chomsky, *Fateful Triangle: The United States, Israel, and the Palestinians*, rev. ed. (Chicago: Haymarket Books, 2014), 74–75.
21. Isabel Kershner, "Netanyahu Criticizes Kerry over Boycott Remarks," *New York Times*, 2 February 2014.
22. Barak Ravid, "Denmark's Largest Bank Blacklists Israel's Hapoalim over Settlement Construction," *Haaretz*, 1 February 2014. Available online at: http://www.haaretz.com/israel-news/1.571849.
23. Harriet Sherwood, "EU Takes Tougher Stance on Israeli Settlements," *Guardian*, 16 July 2013. Available online at: https://www.theguardian.com /world/2013/jul/16/eu-israel-settlement-exclusion-clause.
24. For more on the academic, cultural, and athletic boycott movement, see Dennis Brutus, *Poetry and Protest: A Dennis Brutus Reader*, edited by Aisha Karim and Lee Sustar (Chicago: Haymarket Books, 2006).
25. United Nations Security Council Resolution 418, 4 November 1977. Available online at: http://www.un.org/en/sc/documents/resolutions/1977 .shtml.
26. American Studies Association, Council Resolution on Boycott of Israeli Institutions, 4 December 2013. Available online at: http://www.theasa .net/american_studies_association_resolution_on_academic_boycott_of _israel.
27. Howard Zinn, *You Can't Be Neutral on a Moving Train: A Personal History of Our Times* (1994; repr., Boston: Beacon Press, 2002), 208.
28. Pamela K. Starr, "Mexico's Problematic Reforms," *Current History* 113, no. 760 (February 2014): 51–56.

4. ISIS, THE KURDS, AND TURKEY

1. Ezgi Basaran interview with Graham Fuller, "Former CIA Officer Says US Helped Create IS," *Al-Monitor*, 2 September 2014. Available online at: http://www.al-monitor.com/pulse/politics/2014/09/turkey-usa-iraq -syria-isis-fuller.html.
2. On the sanctions as "genocide," see Denis Halliday interview with David Edwards, Media Lens, May 2000, http://www.medialens.org/index.php /alerts/interviews/77-an-interview-with-denis-halliday.html. On Hans von Sponeck's resignation, see Ewen MacAskill, "Second Official Quits UN Iraq Team," *Guardian*, 15 February 2000, https://www.theguardian .com/world/2000/feb/16/iraq.unitednations.
3. Noam Chomsky and Edward S. Herman, *Manufacturing Consent: The Political Economy of Mass Media* (1988; repr., New York: Pantheon, 2002), 37.

4. This quote was a response to the House Intelligence Committee asking how Kissinger could justify the betrayal of the Kurds; see Daniel Schorr, "Telling It Like It Is: Kissinger and the Kurds," *Christian Science Monitor*, 18 October 1996.

5. Joost R. Hiltermann and *International Herald Tribune*, "Halabja: America Didn't Seem to Mind Poison Gas," *New York Times*, 17 January 2003. Available online at: http://www.nytimes.com/2003/01/17/opinion/halabja -america-didnt-seem-to-mind-poison-gas.html?mcubz=1.

6. Douglas Frantz and Murray Waas, "U.S. Loans Indirectly Financed Iraq Military," *Los Angeles Times*, 25 February 1992. Available online at: http://articles.latimes.com/1992-02-25/news/mn-2628_1_foreign -policy/3.

7. Noam Chomsky, *Deterring Democracy* (London: Verso, 1991), ch. 6.

8. Micah Zenko, *Between Threats and War: U.S. Discrete Military Operations in the Post–Cold War World* (Palo Alto: Stanford University Press, 2010), ch. 3.

9. Jake Hess, interview on *Democracy Now!*, 23 August 2010, transcript available online at: https://www.democracynow.org/2010/8/23/exclusive_us _journalist_deported_from_turkey.

10. Tamar Gabelnick, William D. Hartung, and Jennifer Washburn, "Arming Repression: U.S. Arms Sales to Turkey During the Clinton Administration," Joint Report of the World Policy Institute and the Federation of American Scientists, October 1999. Available online at: https://fas.org /asmp/library/reports/turkeyrep.htm#arms.

11. See table I ibid.

12. "Unperson" is a Newspeak term from George Orwell's *1984* (original ed.: London: Harvill Secker, 1949).

13. For example, see Mohammed Omer, "Gaza's Shejaiya Carnage at Shifa Morgue," *Middle East Eye*, 20 July 2014. Available online at: http://www .middleeasteye.net/news/gazas-shejaiya-carnage-fills-shifa-morgue -1250285665.

14. Keith Wallis, "Kurdish Oil Cargo Unloaded at Sea, Destination a Mystery," Reuters, 21 July 2014. Available online at: http://www.reuters.com /article/us-iraq-oil-kurdistan-asia-idUSKBN0G019720140731/.

15. Sharon Behn, "Unpaid Kurdish Fighters Sign of Economic Woes," *Voice of American News*, 3 September 2014. Available online at: http://www .voanews.com/a/hard-pressed-kurdistan-cannot-pay-security-forces /2944122.html.

16. Nicholas Kristof, "Iran's Proposal for a 'Grand Bargain,'" *New York Times*, 28 April 2007. Available online at: https://kristof.blogs.nytimes.com/2007/04 /28/irans-proposal-for-a-grand-bargain/.

17. Prashad interviewed by Jessica Desvarieux for the Real News Network, 25 September 2014. Transcript available online at: http://www.truth-out .org/news/item/26430-turkish-syrian-border-a-great-hole-in-obama-s -un-speech.

18. "Turkey Rejects U.S. Troop Proposal," CNN, 1 March 2003. Available online at: http://www.cnn.com/2003/WORLD/meast/03/01/sprj.irq .main/.

19. David Ignatius, "A War of Choice, and One Who Chose It," *Washington Post*, 2 November 2003, B1.

20. Catherine Collins, "U.S., Turkey Can't Get over Iraq Question," *Chicago Tribune*, 26 May 2003.

21. Steven Erlanger and Rachel Donadio, "Greek Premier Pledges Vote in December on Debt Deal," *New York Times*, 2 November 2011. Available online at: http://www.nytimes.com/2011/11/03/world/europe/greek -cabinet-backs-call-for-referendum-on-debt-crisis.html.

5. LIVING MEMORY

1. Chris Mooney, "The Arctic Climate Threat That Nobody's Even Talking About Yet," *Washington Post*, 1 April 2015. Available online at: https:// www.washingtonpost.com/news/energy-environment/wp/2015/04/01 /the-arctic-climate-threat-that-nobodys-even-talking-about-yet/?utm _term=.6577644e7a68.

2. Carl Zimmer, "Ocean Life Faces Mass Extinction, Broad Study Says," *New York Times*, 15 January 2015. Alex Morales, "2014 Was World's Hottest Year Since Record Keeping Began in 1880, UN Says," *Bloomberg News*, 2 February 2015. See also Justin Gillis, "2014 Breaks Heat Record, Challenging Global Warming Skeptics," *New York Times*; available online at: http://www.nytimes.com/2015/01/17/science/earth/2014-was-hottest -year-on-record-surpassing-2010.html.

3. PricewaterhouseCoopers, *18th Annual Global CEO Survey 2015: A Marketplace Without Boundaries? Responding to Disruption*, released 13 January 2015. Available online at: https://www.pwc.com/gx/en/ceo-survey/2015/assets /pwc-18th-annual-global-ceo-survey-jan-2015.pdf.

4. Quotation from Davi Kopenawa's autobiography *The Falling Sky* and a series of interviews, excerpted by John Vidal, " 'People in the West Live Squeezed Together, Frenzied as Wasps in the Nest': An Indigenous Yanomami Leader and Shaman from Brazil Shares His Views on Wealth, the Environment and Politics," *Guardian*, 30 December 2014.

5. John Vidal, "Bolivia Enshrines Natural World's Rights with Equal Status for Mother Earth," *Guardian*, 10 April 2011. Available online at: https://

www.theguardian.com/environment/2011/apr/10/bolivia-enshrines
-natural-worlds-rights.

6. Noam Chomsky, *Power Systems: Conversations on Global Democratic Uprisings and the New Challenges to U.S. Empire* (New York: Metropolitan Books, 2013), 161.

7. Arthur Schlesinger, *Robert Kennedy and His Times*, vol. 1 (1978; repr., New York: Mariner/Houghton Mifflin, 2002), 480.

8. Keith Bolender, *Voices from the Other Side: An Oral History of Terrorism Against Cuba* (London: Pluto, 2010).

9. Noam Chomsky, *At War with Asia* (Oakland: AK Press, 2004).

10. Fred Branfman, ed., *Voices from the Plain of Jars: Life Under an Air War*, 2nd ed. (Madison: University of Wisconsin Press, 2013). Introduction available online at: https://zinnedproject.org/materials/voices-from-the
-plain-of-jars/.

11. Fred Branfman, "When Chomsky Wept," *Salon.com*, 17 June 2012, http://
www.salon.com/2012/06/17/when_chomsky_wept/.

12. Alfred W. McCoy, "Foreword: Reflections on History's Largest Air War," in Branfman, *Voices from the Plain of Jars*, xiii.

13. The "battle lab" terminology was used by major generals in Guantánamo and revealed by the commander of the army's Criminal Investigative Task Force; see Mark P. Denbeaux, Jonathan Hafetz, Joshua Denbeaux, et al., "Guantanamo: America's Battle Lab," Center for Policy and Research, Seton Hall University School of Law, January 2015. Report available online at: https://law.shu.edu/policy-research/upload/guantanamo-americas
-battle-lab-january-2015.pdf.

14. Senate Select Committee on Intelligence, *Committee Study of the Central Intelligence Agency's Detention Interrogation Program*, released on 3 April 2014. Available online at: http://www.nytimes.com/interactive/2014/12/09
/world/cia-torture-report-document.html.

15. Floyd Abrams, letter to the editor, *New York Times*, 8 January 2015.

16. See Noam Chomsky, *Who Rules the World?* (New York: Metropolitan Books, 2016), 209.

17. Richard A. Oppel Jr., "Early Target of Offensive Is a Hospital," *New York Times*, 8 November 2004. The photograph that appeared in the print edition was by Shawn Baldwin for the *New York Times*.

18. See Zach Campbell, "Spain Is Sending This Basque Activist to Jail for Writing an Op-Ed," *New Republic*, 11 November 2014. Available online at: https://newrepublic.com/article/120216/spanish-journalist-julen-orbea
-awaits-prison-sentence-eta-article. See also Noam Chomsky, *Necessary Illusions*, Appendix V.7.

19. Kevin Sullivan, "Flogging Case in Saudi Arabia Is Just One Sign of a New Crackdown on Rights Activists," *Washington Post*, 21 January 2015. Available online at: https://www.washingtonpost.com/world/middle_east/a -flogging-in-saudi-arabia-is-just-one-sign-of-a-new-crackdown-on-rights -activists/2015/01/20/e9c50f86-9da0-11e4-86a3-1b56f64925f6_story.html ?utm_term=.d7bf18f53448.

20. Andrew Higgins and Dan Bilefsky, "French Police Storm Hostage Sites, Killing Gunmen," *New York Times*, 9 January 2015, A1. Available online at: https://www.nytimes.com/2015/01/10/world/europe/charlie-hebdo -paris-shooting.html.

21. Andrew Shaver, "You're More Likely to Be Fatally Crushed by Furniture Than Killed by a Terrorist," *Washington Post*, 23 November 2015. Available online at: https://www.washingtonpost.com/news/monkey-cage /wp/2015/11/23/youre-more-likely-to-be-fatally-crushed-by-furniture -than-killed-by-a-terrorist/?utm_term=.51dc230bba8a.

22. Steven Simon, "What Was Behind Israel's Strike in Syria That Killed an Iranian General?" Reuters, 23 January 2015. Available online at: http:// blogs.reuters.com/great-debate/2015/01/23/what-was-behind-israels -strike-in-syria/.

23. Robert Fisk, "Charlie Hebdo: Paris Attack Brothers' Campaign of Terror Can Be Traced Back to Algeria in 1954," *Independent* (London), 9 January 2015.

6. FEARMONGERING

1. Senator Tom Cotton, interviewed by Bob Schieffer, *Face the Nation*, CBS News, 15 March 2015. Transcript available online at: http://www .cbsnews.com/news/face-the-nation-transcripts-march-15-2015-kerry -cotton-manchin/.

2. Hunter Schwarz, "Who Is Tom Cotton?" *Washington Post*, 4 November 2014. Available online at: https://www.washingtonpost.com/news /post-politics/wp/2014/11/04/who-is-tom-cotton/.

3. Mairav Zonszein, "Binyamin Netanyahu: 'Arab Voters Are Heading to the Polling Stations in Droves,'" *Guardian*, 17 March 2015. Available online at: https://www.theguardian.com/world/2015/mar/17/binyamin -netanyahu-israel-arab-election.

4. Amira Hass, interviewed by Amy Goodman, "After Netanyahu Wins Israel Vote with Racism & Vow of Permanent Occupation, How Will World Respond?" *Democracy Now!* 18 March 2015. Available online at: https://www .democracynow.org/2015/3/18/after_netanyahu_wins_israel_vote_with.

5. John Dewey, "The Breakdown of the Old Order," *New Republic*, 25 March 1931: 150–52.

6. Editorial Board, "Packing Guns in the Day Care Center," *New York Times*, 30 November 2015. Available online at: https://www.nytimes.com/2015 /11/30/opinion/packing-guns-in-the-day-care-center.html.

7. Paul Krugman, "Health Care Realities," *New York Times*, 30 July 2009. Available online at: http://www.nytimes.com/2009/07/31/opinion/31krugman .html.

8. Tim Worstall, "It's Not the IMF Demanding Greek Austerity—Quite the Contrary, the Demands Are Not Credible," *Forbes*, 13 December 2016. Available online at: https://www.forbes.com/sites/timworstall/2016/12 /13/its-not-the-imf-demanding-greek-austerity-quite-the-contrary-the -demands-are-not-credible/#480470ef56a8.

9. John Cassidy, "Greece's Debt Burden: The Truth Finally Emerges," *New Yorker*, 3 July 2015. Available online at: http://www.newyorker.com /news/john-cassidy/greeces-debt-burden-the-truth-finally-emerges.

10. Eric Toussaint, "The Cancellation of German Debt in 1953 Versus the Attitude to the Third World and Greece," 18 August 2014. Available online at: http://www.cadtm.org/spip.php?page=imprimer&id_article=10546.

11. David Montgomery, *The Fall of the House of Labor: The Workplace, the State, and American Capitalism* (Cambridge: Cambridge University Press, 1987).

7. ALLIANCES AND CONTROL

1. George Orwell, "Why I Write," *Gangrel* (Summer 1946).

2. Noam Chomsky, *What Kind of Creatures Are We?* (New York: Columbia University Press, 2016).

3. "Autonomous Weapons: An Open Letter from AI and Robotics Researchers," announced July 28, 2015, at the International Joint Conference on Artificial Intelligence. Available online at the Future of Life Institute site: http://futureoflife.org/open-letter-autonomous-weapons/.

4. Kenneth Roth, "The Refugee Crisis That Isn't," *Huffington Post/World-Post*, 3 September 2015, http://www.huffingtonpost.com/kenneth-roth /the-refugee-crisis-that-isnt_b_8079798.html.

5. See, for example, ongoing research on Iraqi refugees done by the Costs of War project at the Watson Institute of International and Public Affairs at Brown University, http://watson.brown.edu/costsofwar/costs/human /refugees/iraqi.

6. Jamey Keaten, "UN Humanitarian Aid Agency: Record $22.2B Needed in 2017," Associated Press, 5 December 2016. Available online at: http://

bigstory.ap.org/article/b595079f0f254ad5b8264d591d934330/un
-humanitarian-aid-agency-222b-funds-needed-2017.

7. Joseph Nevins, "How US Policy in Honduras Set the Stage for Today's Mass Migration," *Conversation*, 1 November 2016. Available online at: http://theconversation.com/how-us-policy-in-honduras-set-the-stage -for-todays-mass-migration-65935.

8. Nick Turse, *Tomorrow's Battlefield: US Proxy Wars and Secret Ops in Africa* (Chicago: Haymarket Books, 2015).

9. Quoted in a December 30, 2009, cable from Clinton when she was the U.S. secretary of state, released by WikiLeaks in December 2010.

10. Claude Moniquet and the European Strategic Intelligence and Security Center (Belgium), "The Involvement of Salafism/Wahhabism in the Support and Supply of Arms to Rebel Groups Around the World," commissioned by the European Parliament's Commission on Foreign Affairs, June 2013. The study is available online at: http://www.europarl.europa .eu/RegData/etudes/etudes/join/2013/457137/EXPO-AFET_ET(2013) 457137_EN.pdf.

11. Mark Curtis, *Secret Affairs: Britain's Collusion with Radical Islam* (London: Serpent's Tail, 2010).

12. Julie Hirschfeld Davis, "Pro-Israel Group Went 'All In,' but Suffered a Stinging Defeat," *New York Times*, 11 September 2015, A1.

13. "US Household Income," Department of Numbers. Available online at: http://www.deptofnumbers.com/income/us/.

14. "Iran, France Sign Agricultural Cooperation Agreement," *Tehran Times*, 22 September 2015. Available online at: http://www.tehrantimes.com /news/249566/Iran-France-sign-agricultural-cooperation-agreement.

15. Jennifer Agiesta, "CNN/ORC Poll: Majority Want Congress to Reject Iran Deal," CNN.com, 28 July 2015. This poll found that 44 percent of Americans approved of the deal and 52 percent said Congress should reject it.

16. Joan McCarter, "Scott Walker Says Maybe He'll Have to Bomb Iran First Day of His Presidency," *Daily Kos*, 20 July 2015. Available online at: http:// www.dailykos.com/story/2015/7/20/1403921/-Scott-Walker-says-maybe -he-ll-have-to-bomb-Iran-first-day-of-his-nbsp-presidency.

17. Noam Chomsky, "The Election, Economy, War, and Peace," *ZNet*, November 25, 2008. Available online at: https://chomsky.info/20081125/.

18. UN News Centre, "United States Vetoes Security Council Resolution on Israeli Settlements," 18 February 2011. Available online at: http://www .un.org/apps/news/story.asp?NewsID=37572#.WOLwUBhh1E4.

19. See Chomsky, *Who Rules the World?*, 140–41, 221.

20. Pope Francis, *Laudato Si': On Care for Our Common Home,* encyclical letter, 24 May 2015. Available online from the Vatican: http://w2.vatican.va /content/dam/francesco/pdf/encyclicals/documents/papa-francesco _20150524_enciclica-laudato-si_en.pdf.

21. Suzanne Goldberg, "Exxon Knew of Climate Change in 1981, but It Funded Deniers for 27 More Years," *Guardian,* 8 July 2015.

22. Bill McKibben, Naomi Klein, and Annie Leonard, "Shell's Arctic Drilling Is the Real Threat to the World, Not Kayaktivists," *Guardian,* 9 June 2015. Available online at: https://www.theguardian.com/environment/2015 /jun/09/shell-oil-greed-undeterred-by-science-climate-change-bill -mckibben-naomi-klein-annie-leonard.

23. Christian is quoted in Chris Isidore and Evan Perez, "GM CEO: 'People Died in Our Cars,'" *CNN Money,* 17 September 2015.

24. For a comprehensive overview of lawsuits against Johnson & Johnson, including pending suits about improperly labeled drugs like Levaquin, see: http://www.johnsonandtoxin.com/lawsuits.shtml.

8. THE ROOTS OF CONFLICTS

1. Noam Chomsky, "The Responsibility of Intellectuals," *New York Review of Books,* 13 February 1967. Available online at: http://www.nybooks .com/articles/1967/02/23/a-special-supplement-the-responsibility-of -intelle/.

2. Noam Chomsky, interviewed by Steven Shalom and Michael Albert, *ZNet,* 30 March 2011.

3. Noam Chomsky, interviewed by C. J. Polychroniou, *Truthout,* 3 December 2015.

4. Andrew Cockburn, *Kill Chain: The Rise of the High-Tech Assassins* (New York: Henry Holt, 2015).

5. Dan Bilefsky and Mark Landler, "As U.N. Backs Military Action in Libya, U.S. Role Is Unclear," *New York Times,* 17 March 2011. Available online at: http://www.nytimes.com/2011/03/18/world/africa/18nations.html.

6. Ian Black, "Bashar al-Assad Implicated in Syria War Crimes, Says UN," *Guardian,* 2 December 2013.

7. Robert Fisk, "David Cameron, There Aren't 70,000 Moderate Fighters in Syria—and Whoever Heard of a Moderate with a Kalashnikov, Anyway?," *Independent* (London), 29 November 2015.

8. Brian Delay, "Indian Polities, Empire, and the History of American Foreign Relations," *Diplomatic History* 39, no. 5 (November 2015): 927–42.

9. Richard W. Van Alstyne, *The Rising American Empire* (New York: Norton, 1960).

10. The United Nations Human Development Index is available online at: http://hdr.undp.org/en/data.

11. Keith Bradsher, "China's Renminbi Is Approved by I.M.F. as a Main World Currency," *New York Times*, 30 November 2015. Available online at: https://www.nytimes.com/2015/12/01/business/international/china -renminbi-reserve-currency.html.

12. Motoko Rich, "Japan Vote Strengthens Shinzo Abe's Goal to Change Constitution," *New York Times*, 10 July 2016. Available online at: https://www .nytimes.com/2016/07/11/world/asia/japan-vote-parliamentary -elections.html.

13. "Protests Erupt as Work Resumes on Futenma Air Base Replacement in Okinawa," *Japan Times*, 6 February 2017. Available online at: http://www .japantimes.co.jp/news/2017/02/06/national/protests-erupt-work -resumes-futenma-air-base-replacement-okinawa/#.WUbvXcaZNE4.

14. John Mitchell, " 'Seconds Away from Midnight': U.S. Nuclear Missile Pioneers on Okinawa Break Fifty Year Silence on a Hidden Nuclear Crisis of 1962," *The Asia-Pacific Journal*, vol. 10, issue 29, no. 1, July 16, 2012. Available online at: http://japanfocus.org/-Jon-Mitchell/3798.

15. Noam Chomsky, *Middle East Illusions: Including Peace in the Middle East? Reflections on Justice and Nationhood* (Lanham, MD: Rowman & Littlefield, 2004).

9. TOWARD A BETTER SOCIETY

1. Merriam-Webster, "Gallery: Word of the Year 2015." Available online at: https://www.merriam-webster.com/words-at-play/word-of-the-year -2015. Catherine Rampell, "Millennials Have a Higher Opinion of Socialism Than of Capitalism," *Washington Post*, 5 February 2016. Available online at: https://www.washingtonpost.com/news/rampage/wp /2016/02/05/millennials-have-a-higher-opinion-of-socialism-than-of -capitalism/.

2. Statement by Hugo Chávez at 61st United Nations General Assembly, 20 September 2006. Available online at: http://www.un.org/webcast/ga/61 /pdfs/venezuela-e.pdf.

3. Mark Tran, "Indian Student Leader Accused of Sedition 'Beaten Up by Lawyers,' " *Guardian*, 17 February 2016.

4. Arundhati Roy, et al., *The Hanging of Afzal Guru and the Strange Case of the Attack on the Indian Parliament*, rev. ed. (Delhi: Penguin India, 2016).

5. BBC News, "India Student Leader Held on Sedition Charges," 12 February 2016. Available online at: http://www.bbc.com/news/world-asia-india -35560518.

6. David Barstow and Suhasini Raj, "Indian Muslim, Accused of Stealing a Cow, Is Beaten to Death by a Hindu Mob," *New York Times*, 4 November 2015. Available online at: https://www.nytimes.com/2015/11/05 /world/asia/hindu-mob-kills-another-indian-muslim-accused-of -harming-cows.html.

7. Leonard Weiss, "What Do Past Nonproliferation Failures Say About the Iran Nuclear Agreement?" *Bulletin of the Atomic Scientists*, 1 September 2015. Available online at: http://thebulletin.org/what-do-past-nonproliferation -failures-say-about-iran-nuclear-agreement8706.

8. Matthew Weaver, "Chomsky Hits Back at Erdoğan, Accusing Him of Double Standards on Terrorism," *Guardian*, 14 January 2016. Available online at: https://www.theguardian.com/us-news/2016/jan/14/chomsky-hits -back-erdogan-double-standards-terrorism-bomb-istanbul.

9. Tuvan Gumrukcu and Orhan Coskun, "Turkey's Erdogan Blames Kurdish Militants After Bomb Kills at Least 13, Wounds 56," Reuters, 17 December 2016. Available online at: http://www.reuters.com/article/us-turkey-blast -idUSKBN14605H.

10. Human Rights Watch, "Turkey: Mounting Security Operation Deaths," 22 December 2015. Available online at: https://www.hrw.org/news/2015 /12/22/turkey-mounting-security-operation-deaths.

11. Agence France-Presse, "Turkish Journalists Charged over Claim That Secret Services Armed Syrian Rebels," *Guardian*, 26 November 2015. Available online at: https://www.theguardian.com/world/2015/nov/27 /turkish-journalists-charged-over-claim-that-secret-services-armed -syrian-rebels.

12. Reporters Without Borders, "RSF Launches International Appeal for Release of Cumhuriyet Journalists," 1 December 2015. Available online at: https://rsf.org/en/news/rsf-launches-international-appeal-release -cumhuriyet-journalists.

13. Noam Chomsky and Christophe Deloire, "Turkey Continues to Muzzle Democracy's Watchdogs," *Washington Post*, 12 November 2015. Available online at: https://www.washingtonpost.com/opinions/turkey-muzzles -democracys-watchdogs/2015/11/12/09c55400-895d-11e5-be8b -1ae2e4f50f76_story.html?utm_term=.1d13e370ae26.

14. Wes Enzinna, "A Dream of Secular Utopia in ISIS' Backyard," *New York Times Magazine*, 24 November 2015. Available online at: https://www .nytimes.com/2015/11/29/magazine/a-dream-of-utopia-in-hell.html.

15. Thomas Ferguson, *Golden Rule: The Investment Theory of Party Competition and the Logic of Money-Driven Political Systems* (Chicago: University of Chicago Press, 1995).

16. Lee Fang, "Gerrymandering Rigged the 2014 Elections for GOP Advan-

tage," BillMoyers.com, 5 November 2016. Available online at: http://billmoyers.com/2014/11/05/gerrymandering-rigged-2014-elections-republican-advantage/.

17. Adam Liptak, "Supreme Court Invalidates Key Part of Voting Rights Act," *New York Times*, 25 June 2013. Available online at: http://www.nytimes.com/2013/06/26/us/supreme-court-ruling.html.

18. Rory McVeigh, David Cunningham, and Justin Farrell, "Political Polarization as a Social Movement Outcome: 1960s Klan Activism and Its Enduring Impact on Political Realignment in Southern Counties, 1960–2000," *American Sociological Review* 79, no. 6 (2014): 1144–71. Available online at: http://www.brandeis.edu/now/2014/december/cunningham-kkk-impact.html.

19. Walter Dean Burnham, "The Changing Shape of the American Political Universe," *American Political Science Review* 59, no. 1 (March 1965): 7–28.

20. Noam Chomsky, "2016 Harvard Trade Union Program," 22 January 2016. Transcript and audio available from Alternative Radio: https://www.alternativeradio.org/products/chon247.

21. Jack Shenker, *The Egyptians: A Radical Story* (New York: New Press, 2016).

22. Declan Walsh, "A Mercedes Shortage? Egypt's Economic Crisis Hits the Rich," *New York Times*, 11 March 2016.

23. Neil MacFarquhar and Merna Thomas, "Russian Airliner Crashes in Egypt, Killing 224," *New York Times*, 31 October 2015. Available online at: https://www.nytimes.com/2015/11/01/world/middleeast/russian-plane-crashes-in-egypt-sinai-peninsula.html.

24. Noam Chomsky, *What Kind of Creatures Are We?* (New York: Columbia University Press, 2015).

25. Noam Chomsky interviewed by David Barsamian, "The Multiple Crises of Neoliberal Capitalism and the Need for a Global Working Class Response," *International Socialist Review* 101 (Summer 2016). Available online at: http://isreview.org/issue/101/multiple-crises-neoliberal-capitalism-and-need-global-working-class-response.

10. ELECTIONS AND VOTING

1. Ben Geman, "Ohio Gov. Kasich Concerned by Climate Change, but Won't 'Apologize' for Coal," *Hill*, 2 May 2012. Available online at: http://thehill.com/policy/energy-environment/225073-kasich-touts-climate-belief-but-wont-apologize-for-coal.

2. Anthony DiMaggio, "Donald Trump and the Myth of Economic Populism: Demolishing a False Narrative," *CounterPunch*, 16 August 2016. Available online at: http://www.counterpunch.org/2016/08/16/donald

-trump-and-the-myth-of-economic-populism-demolishing-a-false-narrative/. See also Anthony DiMaggio, *The Rise of the Tea Party: Political Discontent and Corporate Media in the Age of Obama* (New York: Monthly Review Press, 2011), and Paul Street and Anthony DiMaggio, *Crashing the Tea Party: Mass Media and the Campaign to Remake American Politics* (New York: Routledge, 2011).

3. Drew DeSilver, "For Most Workers, Real Wages Have Barely Budged for Decades," Pew Research Center, 9 October 2014. Available online at: http://www.pewresearch.org/fact-tank/2014/10/09/for-most-workers-real-wages-have-barely-budged-for-decades/.

4. Ryan Teague Beckwith, "Read Hillary Clinton and Donald Trump's Remarks at a Military Forum," *Time*, 7 September 2016. Available online at: http://time.com/4483355/commander-chief-forum-clinton-trump-intrepid/.

5. Thomas Frank interviewed by David Barsamian, "What's the Matter with the Democratic Party?," Alternative Radio, 25 June 2016. Transcript and audio available from Alternative Radio: https://www.alternativeradio.org/collections/spk_thomas-frank/products/frat005.

6. Daniel White, "Read Hillary Clinton's Speech Touting 'American Exceptionalism,'" *Time*, 31 August 2016. Available online at: http://time.com/4474619/read-hillary-clinton-american-legion-speech/.

7. Bob Herbert, "In America: War Games," *New York Times*, 22 February 1998. Available online at: http://www.nytimes.com/1998/02/22/opinion/in-america-war-games.html.

8. Samantha Power, "US Diplomacy: Realism and Reality," *New York Review of Books* 63, no. 13 (18 August 2016). Available online at: http://www.nybooks.com/articles/2016/08/18/us-diplomacy-realism-and-reality/.

9. Michael Smith and Jeffrey M. Jones, "U.S. Satisfaction Remains Low Leading Up to Election," Gallup, 13 October 2016. Available online at: http://www.gallup.com/poll/196388/satisfaction-remains-low-leading-election.aspx.

10. Frank Newport, "As Debate Looms, Voters Still Distrust Clinton and Trump," Gallup, 23 September 2013. Available online at: http://www.gallup.com/poll/195755/debate-looms-voters-distrust-clinton-trump.aspx.

11. Frank Newport, "Congressional Approval Sinks to Record Low," Gallup, 12 November 2016. Available online at: http://www.gallup.com/poll/165809/congressional-approval-sinks-record-low.aspx.

12. Pew Research Center, "Beyond Distrust: How Americans View Their Government," 23 November 2015. Available online at: http://www.people-press.org/2015/11/23/beyond-distrust-how-americans-view-their-government/.

13. BBC News, "Trump Says Putin 'A Leader Far More Than Our President,'" 8 September 2016. Available online at: http://www.bbc.com/news/election -us-2016-37303057.

14. Elise Gould, "U.S. Lags Behind Peer Countries in Mobility," Economic Policy Institute, 10 October 2012. Available online at: http://www.epi.org /publication/usa-lags-peer-countries-mobility/. David Leonhardt, "The American Dream, Quantified at Last," *New York Times*, 8 December 2016. Available online at: https://www.nytimes.com/2016/12/08/opinion/the -american-dream-quantified-at-last.html.

15. John Kenneth Galbraith, *The Affluent Society* (1958; rpt., New York: Mariner Books, 1998), 191.

16. Council of Economic Advisers, "Investing in Higher Education: Benefits, Challenges, and the State of Student Debt," report for the Executive Office of the President, July 2016. Available online at: https://www .whitehouse.gov/sites/default/files/page/files/20160718_cea_student _debt.pdf.

17. Dwight D. Eisenhower, letter to Edgar Newton Eisenhower, 8 November 1954. Available online at: http://teachingamericanhistory.org/library /document/letter-to-edgar-newton-eisenhower/.

18. Alison Smale, "Austria Rejects Far-Right Presidential Candidate Norbert Hofer," *New York Times*, 4 December 2016. Available online at: https:// www.nytimes.com/2016/12/04/world/europe/norbert-hofer-austria -election.html.

19. Molly Moore, "In France, Prisons Filled with Muslims," *Washington Post*, 29 April 2008. Available online at: http://www.washingtonpost.com/wp -dyn/content/article/2008/04/28/AR2008042802560.html.

20. Mark Weisbrot, *Failed: What the Global "Experts" Got Wrong About the Global Economy* (New York: Oxford University Press, 2015).

11. CRISES AND ORGANIZING

1. Image available online at: http://www.spiegel.de/spiegel/print/index -2016-46.html.

2. World Meteorological Organization, *The Global Climate in 2011–2015*, WMO-No. 1179. Available online at: http://library.wmo.int/opac/doc _num.php?explnum_id=3103.

3. United Nations News Centre, "Past Five Years Hottest on Record, Says UN Weather Agency," 8 November 2016. Available online at: http://www .un.org/apps/news/story.asp?NewsID=55503#.WFhHtpKkdRk.

4. John Vidal, " 'There's No Plan B': Climate Change Scientists Fear Consequence of Trump Victory," *Guardian*, 12 November 2016. Available online

at: https://www.theguardian.com/environment/2016/nov/12/climate
-change-marrakech-no-plan-b--trump-victory.

5. Gardiner Harris, "Borrowed Time on Disappearing Land," *New York Times*, 28 March 2014. Available online at: https://www.nytimes.com/2014 /03/29/world/asia/facing-rising-seas-bangladesh-confronts-the -consequences-of-climate-change.html.

6. Pope Francis, "Migrants and Refugees Challenge Us: The Response of the Gospel of Mercy," 17 January 2016. Available online at: https://w2 .vatican.va/content/francesco/en/messages/migration/documents/papa -francesco_20150912_world-migrants-day-2016.html.

7. United Nations High Commissioner for Refugees, *Global Trends: Forced Displacement in 2015*. Available online at: http://www.unhcr.org/576408cd7.pdf.

8. Ben Dreyfuss, "Germany Has Taken In 800,000 Refugees. Guess How Many the US Has Taken In?" *Mother Jones*, 3 September 2015. Available online at: http://www.motherjones.com/mojo/2015/09/germany-has -taken-800000-refugees-guess-how-many-us-has-taken.

9. Vince Chadwick, "Erdoğan Slams 'Hypocritical' EU over Human Rights Criticism," *Politico* (European edition), 21 March 2016. Available online at: http://www.politico.eu/article/recep-tayyip-erdogan-slams-hypocri tical-eu-over-human-rights-criticism-turkey-migration-crisis-europe -refugees/.

10. WaterAid released a report in March 2016 for World Water Day. See Bihuti Agarwal, "Indians Have the Worst Access to Safe Drinking Water in the World," *Wall Street Journal*, 22 March 2016.

11. Andy Barr, "The GOP's No-Compromise Pledge," *Politico*, 28 October 2010. Available online at: http://www.politico.com/story/2010/10/the-gops -no-compromise-pledge-044311.

12. Donald Trump, 6 December 2016, Fayetteville, NC. See Steve Holland, "Trump Lays Out Non-Interventionist Military Policy," Reuters, 6 December 2016. Available online at: http://www.reuters.com/article/us-usa -trump-military-idUSKBN13W06L.

13. Jonathan Weisman, "Reagan Policies Gave Green Light to Red Ink," *Washington Post*, 9 June 2004, A11. Available online at: http://www.washington post.com/wp-dyn/articles/A26402-2004Jun8.html.

14. Patti Domm, "Peabody Energy Shares Rocket After Trump Wins Presidency," Market Insider, CNBC, 9 November 2016, http://www.cnbc.com /2016/11/09/peabody-energy-shares-rocket-after-trump-wins-presidency .html.

15. Lawrence Mishel, Elise Gould, and Josh Bivens, "Wage Stagnation in Nine Charts," Economic Policy Institute, 6 January 2015. Available online at: http://www.epi.org/publication/charting-wage-stagnation/.

16. Arlie Hochschild, *Strangers in Their Own Land: Anger and Mourning on the American Right* (New York: New Press, 2016).

17. The Omaha People's Party Statement is collected in Howard Zinn and Anthony Arnove, eds., *Voices of a People's History of the United States* (New York: Seven Stories, 2004).

18. Brady Dennis and Steven Mufson, "As Trump Administration Grants Approval for Keystone XL Pipeline, an Old Fight Is Reignited," *Washington Post*, 24 March 2017. Available online at: https://www.washingtonpost.com/news/energy-environment/wp/2017/03/24/trump-administration-grants-approval-for-keystone-xl-pipeline/?utm_term=.dabc79d21200.

12. THE TRUMP PRESIDENCY

1. Kevin Liptak and Dan Merica, "Trump Believes Millions Voted Illegally, WH Says—but Provides No Proof," CNN, 25 January 2017. Available online at: http://www.cnn.com/2017/01/24/politics/wh-trump-believes-millions-voted-illegally/index.html.

2. Philip Rucker, Ellen Nakashima, and Robert Costa, "Trump, Citing No Evidence, Accuses Obama of 'Nixon/Watergate' Plot to Wiretap Trump Tower," *Washington Post*, 14 March 2017. Available online at: https://www.washingtonpost.com/news/post-politics/wp/2017/03/04/trump-accuses-obama-of-nixonwatergate-plot-to-wire-tap-trump-tower/.

3. Alan Rappeport, "Bill to Erase Some Dodd-Frank Banking Rules Passes in House," *New York Times*, 8 June 2017. Available online at: https://www.nytimes.com/2017/06/08/business/dealbook/house-financial-regulations-dodd-frank.html.

4. Thomas E. Mann and Norman J. Ornstein, "Finding the Common Good in an Era of Dysfunctional Governance," *Dædalus: The Journal of the American Academy of Arts & Sciences* 142, no. 2 (Spring 2013).

5. Hans M. Kristensen, Matthew McKinzie, and Theodore A. Postol, "How US Nuclear Force Modernization Is Undermining Strategic Stability: The Burst-Height Compensating Super-Fuze," *Bulletin of the Atomic Scientists*, 1 March 2017. Available online at: http://thebulletin.org/how-us-nuclear-force-modernization-undermining-strategic-stability-burst-height-compensating-super10578.

6. Jonathan Easley, "Poll: Bernie Sanders Country's Most Popular Active Politician," *Hill*, 18 April 2017. Available online at: http://thehill.com/homenews/campaign/329404-poll-bernie-sanders-countrys-most-popular-active-politician.

7. Robert Pollin, *Greening the Global Economy* (Cambridge, MA: MIT Press, 1995).

8. Gar Alperovitz, *America Beyond Capitalism: Reclaiming Our Wealth, Our Liberty, and Our Democracy*, 2nd ed. (Washington, DC: Democracy Collaborative Press and Dollars and Sense, 2011).

9. Karl Marx, *The Eighteenth Brumaire of Louis Napoleon* (New York: International Publishers, 1994), ch. 7.

10. Gary Milhollin, testimony in *United States Export Policy Toward Iraq Prior to Iraq's Invasion of Kuwait: Hearing Before the Committee on Banking, Housing, and Urban Affairs*, U.S. Senate, 102nd Congress, 27 October 1992. See also Chomsky, *Hegemony or Survival* (New York: Metropolitan Books, 2003), 111–12, and Chomsky, *Failed States* (New York: Metropolitan Books, 2006), 28–29.

11. David E. Sanger and Gardiner Harris, "U.S. Pressed to Pursue Deal to Freeze North Korea Missile Tests," *New York Times*, 21 June 2017. Available online at: https://www.nytimes.com/2017/06/21/world/asia/north-korea -missle-tests.html.

12. Blaine Harden, "The U.S. War Crime North Korea Won't Forget," *Washington Post*, 24 March 2015. Available online at: https://www.washingtonpost .com/opinions/the-us-war-crime-north-korea-wont-forget/2015/03/20 /fb525694-ce80-11e4-8c54-ffb5ba6f2f69_story.html.

13. For further analysis, see Noam Chomsky, *The Essential Chomsky*, ed. Anthony Arnove (New York: New Press, 2008), 185–86.

14. Choe Sang-Hun, "South Korea Elects Moon Jae-in, Who Backs Talks with North, as President," *New York Times*, 9 May 2017. Available online at: https://www.nytimes.com/2017/05/09/world/asia/south-korea-election -president-moon-jae-in.html.

15. See the DiEM25 website at: https://diem25.org/.

16. Peter Holley, Abby Ohlheiser, and Amy B. Wang, "The Doomsday Clock Just Advanced, 'Thanks to Trump': It's Now Just 2½ Minutes to 'Midnight,'" *Washington Post*, 26 January 2017. Available online at: https:// www.washingtonpost.com/news/speaking-of-science/wp/2017/01/26 /the-doomsday-clock-just-moved-again-its-now-two-and-a-half -minutes-to-midnight/.

INDEX

Abe, Shinzō, 128
abortion, 181
Abrams, Floyd, 83
Abu Ghraib, 87
Abunimah, Ali, 92
activism, 30, 37
Adivasis, 28
advertising, 13, 15–16, 32–33
Affordable Care Act (Obamacare,
　　ACA), 93–94
Afghanistan, 7, 66, 101, 119
Afghan refugees, 101, 169
Africa, 73, 102–3, 127
African Americans, 55, 144–45, 176
African National Congress
　　(ANC), 50
Afro-Colombians, 77
Ahab, King, 46
Ahrar al-Sham, 122
AIG, 40

Ai Weiwei, 53
al-Anfal campaign, 62
Albright, Madeleine, 156
Algeria, 87–88
Alinsky, Saul, 30
Al Nusra Front, 66, 122
Alperovitz, Gar, 184
Al Qaeda, 7, 83, 119, 122
Amazon, 76
American Dream, 176
American exceptionalism, 156
American Israel Public Affairs
　　Committee (AIPAC), 25, 105, 107
American Legion National Conven-
　　tion (2016), 156
American Legislative Exchange
　　Council (ALEC), 39–40
American Revolution, 15, 138
American Studies Association, 54
Amnesty International, 63, 188

NOAM CHOMSKY

WHO RULES THE WORLD?

'The closest thing in the English-speaking world to an intellectual superstar' *Guardian*

As long as the general population is passive, apathetic, diverted to consumerism or hatred of the vulnerable, the powerful can do as they please and those who survive will be left to contemplate the outcome . . .

In the post-9/11 era, America's policy-makers have increasingly prioritised the pursuit of power, both military and economic, above all else – human rights, democracy, even security. Drawing on examples ranging from expanding drone assassination programs to civil war in Syria to the continued violence in Iraq, Iran, Afghanistan, Israel and Palestine, philosopher, political commentator and prolific activist Noam Chomsky offers unexpected and nuanced insights into the workings of imperial power in our increasingly chaotic planet.

NOAM CHOMSKY

HOW THE WORLD WORKS

'Chomsky has an authority granted by brilliance' *Sunday Times*

Divided into four sections and originally published in the US as individual short books which have collectively sold over half a million copies, *How the World Works* is a collection of speeches and interviews with Chomsky by David Barsamian, edited by Arthur Naiman. It includes 'What Uncle Sam Really Wants', about US foreign policy; 'The Prosperous Few and the Restless Many', about the new global economy, food, Third World 'economic miracles' and the roots of racism; 'Secrets, Lies and Democracy', about the US, the CIA, religious fundamentalism, global inequality and the coming eco-catastrophe; and 'The Common Good', about equality, freedom, the media, the myth of Third World debt and manufacturing dissent.

With exceptional clarity and power of argument, Noam Chomsky lays bare as no one else can the realities of contemporary geopolitics.

NOAM CHOMSKY

HEGEMONY OR SURVIVAL

'For anyone wanting to find out about the world we live in, there is one simple answer: read Noam Chomsky' *New Statesman*

The United States is in the process of staking out not just the globe, but the last unarmed spot in our neighbourhood – the skies – as a militarized sphere of influence. Our earth and its skies are, for the Bush administration, the final frontiers of imperial control. In *Hegemony or Survival*, Noam Chomsky explains how we came to this moment, what kind of peril we find ourselves in, and why our rulers are willing to jeopardize the future of our species. In our era, Chomsky argues, empire is a recipe for an earthly wasteland.

From the world's foremost intellectual activist comes an irrefutable analysis of America's pursuit of total domination and the catastrophic consequences that are sure to follow.

NOAM CHOMSKY

FAILED STATES

'One of the greatest, most radical public thinkers of our time'
Arundhati Roy

The United States asserts the right to use military force against 'failed states' around the globe. But as Noam Chomsky argues in this devastating analysis, America shares features with many of the regimes it insists are failing and constitute a danger to their neighbours.

Offering a comprehensive and radical examination of America past and present, Chomsky shows how this lone superpower – which topples foreign governments, invades states that threaten its interests and imposes sanctions on regimes it opposes – has stretched its own democratic institutions to breaking point. And how an America in crisis places the world ever closer to the brink of nuclear and environmental disaster.

NOAM CHOMSKY

HOPES AND PROSPECTS

'One of our most valued political thinkers' *Independent on Sunday*

In this urgent new book, Noam Chomsky surveys the threats and prospects of our early twenty-first century. Exploring challenges such as the growing gap between North and South, American exceptionalism (even under Obama), the fiascos of Iraq and Afghanistan, the Israeli assault on Gaza and the recent financial bailouts, he also sees hope for the future and a way to move forward – in the so-called democratic wave in Latin America and in the global solidarity movements which suggest 'real progress towards freedom and justice'.

Hopes and Prospects is essential reading for anyone who is concerned about the primary challenges still facing the human race and is wondering where to find a ray of hope.